MATH HOUSE
Online

Learn By Doing: Algebra I

An Active Approach to Learning Mathematics

Ryan Hobbs

Delaware
Spring Press

ISBN: 978-1-7332514-3-3

This book follows the structure of Marecek, Lynn, et al. *Elementary Algebra 2e*. OpenStax, 2020, which is available for free at https://openstax.org/details/books/elementary-algebra-2e.
The traditional lecture videos, linked by QR code at the end of each lesson, use the material from *Elementary Algebra 2e*, and are available for free at https://www.youtube.com/@rphobbs2002.

The Reason for this Workbook

I have been a college math lecturer for over ten years and it didn't take me long to develop my foremost principle of education—if students aren't awake then they aren't learning. In my opinion, for most math students, the traditional lecture approach has limited value. And so, I began a journey to develop a more interactive approach to the mathematics classroom. This workbook is the result.

However, there is much more here than just a series of worksheets, as I've strived to incorporate the other principles of math education which I've discovered over the years.

Math is a language. Good teaching must translate this language of numbers into English. Language learning also requires "comprehensible input." It is good to be challenged and stretched, but when teaching is beyond a student's level of understanding they become overwhelmed and shut down. As math progresses, it is valuable to show students how more advance concepts relate to simpler concepts which they already know and understand. And, when possible, it helps to make math visual.

Yet perhaps most importantly, good teaching comes beside students when they need help. This is why every activity has a QR code which links to a walk-through video of the active learning lesson.

In other words, good teaching is very much like good tutoring, and that is the fundamental idea which has guided the development of this course.

How to Use this Book

Each activity in this workbook has a QR code in the top righthand corner. This QR code will take you to a YouTube video where I work and explain every problem and concept in that lesson. It is my attempt to use technology to come along side of a student just as I would do in a real classroom.

So, as you work the activity, start and stop the video, as needed, for help. Or, skip through any part of the video which you don't need. The back of the workbook also contains a solution guide for checking your answers.

If after finishing an activity, you feel as if you could use some additional help, a QR code in the bottom right links to a traditional lecture for that section.

This workbook is structured to follow the OpenStax textbook Elementary Algebra. The activity numbers correspond to the chapters and sections of that book. For more practice problems, the textbook is available for free on the OpenStax website.

Table of Contents

Learning to Solve Word Problems

Graphs

Systems of Linear Equations

Polynomials

Factoring

Rational Expressions and Equations

Roots and Radicals

Quadratic Equations

Appendix

We are starting out with something that you may be very comfortable with—the place value system. Basically, we organize numbers by their position.

Trillion			Billion			Million			Thousand			Ones		
Hundred Trillion	Ten Trillion	Trillion	Hundred Billion	Ten Billion	Billion	Hundred Million	Ten Million	Million	Hundred Thousand	Ten Thousand	Thousand	Hundred	Tens	Ones

Each large group has a unique name, but is made up of three smaller groups: Hundreds, Tens and Ones. Any whole number can be placed in their appropriate position.

$$678,023,152$$

Here I have placed them into their position in the chart:

Trillion			Billion			Million			Thousand			Ones		
Hundred Trillion	Ten Trillion	Trillion	Hundred Billion	Ten Billion	Billion	Hundred Million	Ten Million	Million	Hundred Thousand	Ten Thousand	Thousand	Hundred	Tens	Ones
						6	7	8	0	2	3	1	5	2

Give the name of the position for each of the following digits:

a) 8

b) 0

c) 1

d) 7

e) 6

I've given you a blank chart. Put each digit of the following number into its proper place.

701,237,941,002,149

f)

Trillion			Billion			Million			Thousand			Ones		
Hundred Trillion	Ten Trillion	Trillion	Hundred Billion	Ten Billion	Billion	Hundred Million	Ten Million	Million	Hundred Thousand	Ten Thousand	Thousand	Hundred	Tens	Ones

In math, when we write the name of a number, it is important not to use the word "and." We reserve the word "and" for the decimal place. Put the name of the following numbers in words:

g) 174

h) 234

i) 549

For larger numbers, we don't name every place. We write each group as if it were a three-digit number and then add on the group name. Here's an example:

598 162 321

Five hundred ninety-eight *million*.

One hundred sixty-two *thousand*.

Three hundred twenty-one.

j) Name the following number:

174,234,549

Try one more:

k)   621,114,987,002,581

We can, of course, go the other way. Write out the following number. I've added a blank chart if you would like to use it.

One hundred seventy-two billion, Two hundred fifteen million, seven hundred two thousand, three hundred one.

l)

Trillion			Billion			Million			Thousand			Ones		
Hundred Trillion	Ten Trillion	Trillion	Hundred Billion	Ten Billion	Billion	Hundred Million	Ten Million	Million	Hundred Thousand	Ten Thousand	Thousand	Hundred	Tens	Ones

Try another. (Be careful, I've skipped over an entire group. And, watch out for any places which need a zero.)

m) Seven trillion, four hundred six million, three hundred forty-seven thousand, fifteen.

Our final idea for this activity is rounding. Rounding rules are simple. Find the place to the right of the one we want to round. If the number is a 5, 6, 7, 8, or 9 then round up. If the number is a 0, 1, 2, 3, or 4. We leave our number the same and make everything to the right of it a zero. Here are some examples:

Round 5,612 to the thousands place.

We want to round to the thousands. I've circled it.

(5,612

Check the number to the right of the place we want.

(5,612

It is a 6, so we round the 5 up and everything else to the right becomes a zero.

6,000

Round 5,612 to the hundreds place.

5,612

We want to round to the hundreds. I've circled it.

5,6,12

Check the number to the right of the place we want.

5,6,12

It is a 1, so we don't round the 6. It stays as it is and everything else to the right becomes a zero.

5,600

Work some.

n) Round 7,813,555,399 to the millions place.

o) Round 7,813,555,399 to the ten thousands place.

p) Round 7,813,555,399 to the billions place.

q) Round 7,813,555,399 to the hundred thousands place.

I believe that understanding factors is one of the keys to algebra. In this activity, we want to review some key ideas of factoring.

First, numbers can be classified into two categories: prime numbers and composite numbers. Composite numbers are made up of two or more other numbers which are being multiplied together.

$$6 = 2 \cdot 3$$

$$15 = 3 \cdot 5$$

$$30 = 2 \cdot 3 \cdot 5$$

Prime numbers are those which cannot. They can only be written as a product by multiplying themselves by 1.

$$5 = 5 \cdot 1$$

$$7 = 7 \cdot 1$$

$$59 = 59 \cdot 1$$

a) Look at this list of numbers. Circle any primes.

8   13   21   35   17   61   100

Finding the factors which make up a number is a very important algebra skill. In order to find factors, we need to know what numbers can divide into our number. To make that easier, there are divisibility rules. If the number we are dividing fits certain rules, we can find factors faster.

Here is a list of the most common divisibility rules:

- Divisible by 2: If the number is an even number. (The last digit is a 0, 2, 4, 6, or 8.)
- Divisible by 3: If the sum of the digits is divisible by 3.
- Divisible by 5: If the last digit ends in 5 or 0.
- Divisible by 6: If a number is divisible by 2 and 3.
- Divisible by 10: If a number ends in 0.

b) Circle any numbers divisible by 2.

15   16   102   2434   4   16   81   156

c) Circle any number divisible by 3.

For example: 132. Add the digits. $1 + 3 + 2 = 6$. Six is divisible by 3, so 132 is divisible by 3.

231   6033   122   42   63   99   10547

d) Circle any number divisible by 5.

15   103   7855   239670   65   8000   98

e) Circle any numbers divisible by 6.

For example: 462. It is divisible by 2 because it ends in an even number. It is divisible by 3 because $4 + 6 + 2 = 12$ is divisible by 3.

81   240   550   1212   142   6954

f) Circle any numbers divisible by 10.

1000   5020   5025   100365   84750   90   21054780

In algebra, it will often be useful to find the prime factorization of a number. If a number is composite, it is the product of other numbers. For instance, here is the prime factorization of the number 240.

$$240 = 2 \cdot 2 \cdot 2 \cdot 2 \cdot 3 \cdot 5$$

It is important to understand that the number 240 is the same as $2 \cdot 2 \cdot 2 \cdot 2 \cdot 3 \cdot 5$. Finding a prime factorization requires dividing down 240. I use a method called a factor tree.

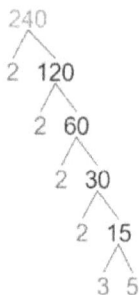

240
2 120
  2 60
    2 30
      2 15
        3 5

Keep dividing the number. When a number is prime—stop.

The prime factorization is the list of all the prime number multiplied together.

$$240 = 2 \cdot 2 \cdot 2 \cdot 2 \cdot 3 \cdot 5$$

Using the division rules is helpful, but I usually just try to divide by 2. If that doesn't work, I try 3. Then, I try 5. Of course, there are more possibilities, but most numbers will depend on those three.

Here is the factor tree for 165:

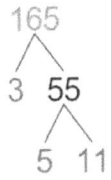

```
    165
    /\
   3  55
      /\
     5  11
```

g) Write the prime factorization:

165:

Here is the factor tree for 2358:

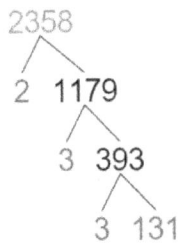

```
   2358
   /\
  2  1179
      /\
     3  393
         /\
        3  131
```

h) Write the prime factorization:

2358:

i) Make the factor tree and give the prime factorization for 81.

j) Make the factor tree and give the prime factorization for 64.

k) Make the factor tree and give the prime factorization for 125.

One of the applications of prime factorizations is to find something called the least common multiple (LCM) for a pair of numbers. First, let's see the most basic way to find the LCM. Suppose we wanted to find the LCM for the numbers 8 and 10. We could, literally, write out all the multiples for each number until we found the lowest number that they both share.

$$8: 8, 16, 24, 32, 40, 48, 56, 64, 72, 80 \ldots$$

$$10: 10, 20, 30, 40, 50, 60, 70, 80 \ldots$$

If we compare the lists, we find that the lowest match if 40. Therefore, the LCM is 40. Notice that there are other multiples. 80 is also a match. However, it is not the lowest.

Here is a list of the multiples for the numbers 12 and 15.

$$12: 12, 24, 36, 48, 60, 72, 84, 96 \ldots$$

$$15: 15, 30, 45, 60, 75, 90 \ldots$$

l) What is the least common multiple?

m) Try this list approach to find the LCM for the numbers 15 and 21.

15:

21:

The list method works, but it can be time consuming, especially with larger numbers. Prime factorizations can make the work much easier.

Here are the prime factorizations for 8 and 10:

$8 = 2 \cdot 2 \cdot 2$

$10 = 2 \cdot 5$

I've circled factors which they share.

$$8 = 2 \cdot 2 \cdot 2$$
$$10 = 2 \cdot 5$$

Now, create a factorization with every factor from both numbers. However, count the shared factors only once.

$2 \cdot 2 \cdot 2 \cdot 5$

The least common multiple is: $2 \cdot 2 \cdot 2 \cdot 5 = 40$

Let's walk through the idea again. This time with 12 and 30.

Here are the prime factorizations.

$12 = 2 \cdot 2 \cdot 3$

$30 = 2 \cdot 5 \cdot 3$

(I wrote 30's factorization out of order just to make the circling easier.)

Now, we circle matching factors.

$$12 = 2 \cdot 2 \cdot 3$$
$$30 = 2 \cdot 5 \cdot 3$$

Counting matches only once, we list out the factors to find the LCM.

$$2 \cdot 2 \cdot 5 \cdot 3 = 60$$

Use the prime factorization method to find the LCM for the following:

n)   Find the LCM of 12 and 15.

o)   Find the LCM for 15 and 21.

p)   Find the LCM for 192 and 168.

Much of chapter one is a collection of terms and ideas designed to set a foundation for our work. Likely, it is a review for you. However, it is very important. The better you understand the terms and concepts; the easier the rest of algebra will be.

The most basic idea in algebra is the variable.

$$x + 3$$

We don't know the value of the number x. It could vary. Thus, the name variable. The 3 isn't varying, so it is called a constant.

In algebra, constants and variables will combine to make one of two things: expressions or equations. Equations have an equal sign and can be solved to find a value for the variable. Expressions don't have an equal sign, and therefore, they can't be solved. The best we can do with an expression is to simplify it. This seems pretty easy, but I will often have students trying to solve expressions. It is a major problem and one to avoid.

a) Here, circle anything which is an equation. Put a square around anything which is an expression.

$$3x - 5$$

$$5y + 15 = 20$$

$$4z^2 - 7z + 18$$

$$\frac{r - 3}{r + 2} = 4$$

$$(x - 2)(x + 5) = 12$$

$$x^3 - 27$$

Exponents are repeated multiplication. For instance,

$$2^3 = 2 \cdot 2 \cdot 2 = 8$$

Give the value of each of the following exponents.

b) $3^4 =$

c) $12^2 =$

d) $5^3 =$

e) $(-2)^3 =$

Later, we will see problems involving inequalities. Variables can take on a range of values with an inequality.

$$x > 4$$

$$y \leq 2$$

Students often forget how the inequalities work. As in reading, math goes left to right. And, the larger opening is always toward the bigger number.

$>$

Greater Than

$<$

Less Than

$\geq$

Greater Than or Equal To

$\leq$

Less Than or Equal To

Math is a language, and learning it is like learning Spanish or Japanese. So, we often need to translate between math language and the English language. Below are a series of math expressions involving inequalities. Match each inequality on the left with the correct translation on the right.

f) $17 > 2$ x is less than or equal to two

g) $x \geq 8$ x is greater than or equal to eight

h) $-5 < 7$ seventeen is greater than two

i) $x \leq 2$ negative five is less than seven

Next, we will do some more translation from math language to English. We begin with addition and subtraction. Match the math expression on the left with its English translation on the right. (There are two right answers for each.)

Six more than x.

j) $x + 6$

Two subtracted from x.

k) $x - 2$ Six added to x.

Two less than x.

Translate each of the following math expressions into English:

l) $x + 15$

m) $y - 7$

Match some harder addition and subtraction translations: (There is only one correct answer per translation.)

n) $4x^2 + 15$ The difference of two x-squared and three y-squared.

o) $17x - 4y$ The difference of seventeen x and four y.

p) $9x^3 + 6$ The sum of four x-squared and fifteen.

q) $2x^2 - 3y^2$ The sum of nine x-cubed and six.

Translate each of the following math expressions into English:

r) $5x^2 + 21$

s) $5x^2 - 21$

We can also do multiplication and division. Match the following:

t) $8x$ The quotient of x-squared and six.

u) $\dfrac{8}{x}$ The product of eight and x.

v) $6x^2$ The quotient of eight and x.

w) $\dfrac{x^2}{6}$ The product of six and x-squared.

Translate each of the following math expressions into English:

x) $12x$

y) $\dfrac{x}{12}$

z) $\dfrac{12}{x}$

The hardest to translate are those involving grouping symbols. It will be a product times a sum. See if you can distinguish a difference as you match the following.

aa) $3(x + y)$ The sum of three times x and three times y.

bb) $3x + 3y$ Two times the difference of r and s.

cc) $2(r - s)$ The difference of two times r and two times s.

dd) $2r - 2s$ Three times the sum of x and y.

Translate each of the following math expressions into English:

ee) $12(a + b)$

ff) $12a - 12b$

15

Algebra I

Active Lesson: 1.2b

Look at the problem below:

$$25 \div 5^2 + 3(10 - 8)$$

Mathematicians realized that without guidelines two different students could tackle this problem in different ways, resulting in different answers. So, a set of instructions were developed, called the Order of Operations. It is like a recipe for how to tackle problems such as this. There is an acronym to help you remember.

Please	Parenthesis (Actually, any grouping symbol.)
Excuse	Exponents
My	Multiplication
Dear	Division
Aunt	Addition
Sally	Subtraction

In truth, multiplication and division are actually tied. Addition and subtraction are also tied. To break a tie, we move left to right across the expression, just like you would do if you were reading. Let's work the original problem now that we have the order in which we are to work.

$$25 \div 5^2 + 3(10 - 8)$$

Parenthesis

$$25 \div 5^2 + 3(2)$$

Exponents

$$25 \div 25 + 3(2)$$

Multiplication/Division

There is both. Remember 3(2) is actually multiplication. So, we'll go left to right.

$$1 + 3(2)$$

$$1 + 6$$

Try some on your own.

a) $6^2 - 3 + 2(3 + 5)$

b) $100 \div (7 + 3) - 2 \cdot 2^2 + 12$

This next set has grouping symbols inside of grouping symbols. When this happens work your way from the inside group outward.

c) $3 + 3^3 + 2[8 - 3(5 - 3)]$

d) $4[15 - 2(3 + 1)] + 6^2 \div 2$

Our next idea is evaluating an expression. In algebra, evaluating means that we are going to replace the unknown value of a variable with a known value. When we do this, it is important that we put the replacement value inside of a parenthesis. It won't always turn out to change the outcome of the problem, but without it, multiplication could get lost. Look at this expression:

$$5x - 3$$

Let $x = 2$. We substitute in parenthesis. (If not, we may lose the fact that we need to do 5 times 2.)

$$5(2) - 3$$

After substituting, just remember the order of operations.

$$10 - 3 = 7$$

Try these:

$$7x + 15$$

e)   Let $x = 3$

f)   Let $x = -3$

When the expressions become more complicated, the order of operations is essential. Look at this expression: (Notice that there are multiple x's that must be replaced.)

$$2x^2 - 3x + 5$$

Let $x = 3$.

$$2(3)^2 - 3(3) + 5$$

Following the order of operations, exponents come first:

$$2(9) - 3(3) + 5$$

Next, multiplication. (We would go left to right here.)

$$18 - 3(3) + 5$$

$$18 - 9 + 5$$

Then addition/subtraction, from left to right.

$$9 + 5$$

$$14$$

Try one on your own:

$$3x^2 - 5x + 21$$

g) Let $x = 4$

h) Let $x = 6$

i) Let $x = -2$

Algebra I

Active Lesson: 1.2c

a) Add the following:

$$5 + 5 + 5 + 5 + 5 + 5$$

b) Repeated addition is the same as multiplication. Which of the following would give the same answer as the problem above?

    a) $6 \cdot 5$
    b) $7 \cdot 5$
    c) $8 \cdot 5$

This works the same with variables.

$$x + x + x + x + x + x$$

c) Which is the same?

    a) $5x$
    b) $7x$
    c) $6x$

Look at the following.

$$2 \cdot 5 + 6 \cdot 5$$

d) Which is the same?

    a) $7 \cdot 5$
    b) $6 \cdot 5$
    c) $8 \cdot 5$

Look at this.

$$3x + 4x$$

e) Which is the same?

    a) $8x$
    b) $6x$
    c) $7x$

Subtraction doesn't change the idea. Simplify the following:

f) $12x - 4x + 6x =$

g) $-3x + 20x - 4x =$

One of the most foundational ideas in algebra is combining like terms. It simply means we can add matching variables. Only x's go with x's and y's with y's. Combine the following like terms:

h) $4x + 6y + 2x + 4y$

i) $14x - 3y + 12x - 2y$

The principal is the same even when things get more complicated. The key is to think of the variable portion as a word. To combine them, the word must be identical. Nothing combines in the following because even though they are similar, they are not the exact same word. One is x^2 and the other is x.

$$4x^2 + 3x$$

However, these have matches. Combine them.

$$4x^2 + 3x + 2x + 5x^2$$

j) Which is the correct answer?

    a) $7x^2 + 7x$

    b) $9x^2 + 5x$

    c) $8x^2 + 6x$

Combine these:

k) $14x^2 - 6x + 5x^2 + 3x - 2x^2$

l) $2x^2 + 7y + 5y^2 + 3x - 3y^2 + 6x^2$

m) $5xy + 6xy$

n) $11x^2y + 3x^2y$

o) $11x^2y + 3xy^2 - 2xy^2 + 10x^2y$

Be sure that you are matching identical "words." It is the key to combining like terms. Watch out for one more thing. If there is no number in front of a variable, it means that there is a one in front.

$$x + 5x = 6x$$

Combine like terms:

p)
$$x^2 + y + 5y^2 + 3y + 2y^2 + 4x^2$$

Finally, look at this expression:

$$14x^2 + 2x + 3$$

There are three terms in this expression: $14x^2$, $2x$, 3. Terms are separated by + or -. (They are numbers or variables being held together by multiplication or division.)

Coefficients are any numbers in front of variables. There is a coefficient of 14 in front of the x^2. And, there is a coefficient of 2 in front of the x.

List the terms and coefficients in this expression. (Hint: The 15 owns the negative sign, so it is -15. The x has a coefficient although it doesn't look like it. The coefficient is 1.)

$$6x^3 - 15x^2 + x + 9$$

q) Term:

r) Coefficients:

Algebra I

Active Lesson: 1.3a

Math works like reading. We go from left to right. So, when we look at a number line, the number increase as we move to the left.

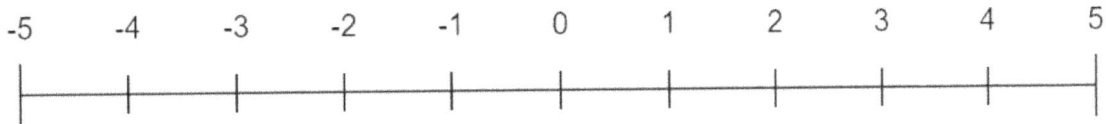

$$-5 \quad -4 \quad -3 \quad -2 \quad -1 \quad 0 \quad 1 \quad 2 \quad 3 \quad 4 \quad 5$$

Negative numbers give students trouble. The best idea which I've found to help is to think of it like money. Negatives represent debt. Therefore, the larger the negative number, the worse the debt problem.

$$-5 < -2$$

Add a > or < sign to each of the following to indicate which is the larger number.

a)   4__7

b)   15__3

c) -2__5

d) -8__-1

e) -9__-11

Another useful idea for negative signs is to think of them as opposites. So, the opposite of 5 would -5. The opposite of -8 would be 8. Answer the following:

f) Give the opposite of 6

g) Give the opposite of 101

h) Give the opposite of -7

i) Give the opposite of -25

j) What does $-(-8)$ equal?

k) What does $-(-21)$ equal?

Math is a language. It is the language of numbers. For the mathematician, math is also a language that simplifies. Although not as familiar today as it once was, math is like shorthand. Shorthand is used to abbreviate the English language to allow faster writing.

Our next idea is called the absolute value. The question a mathematician wanted to know was, "how far away is this number from zero?"

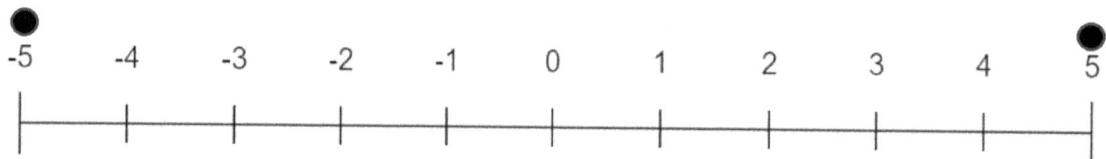

The mathematician didn't want the number. They just wanted the distance to zero. So, both the number 5 and the number -5 are five away from zero.

Basically, an absolute value makes the negative sign meaningless. And because math is like shorthand, the entire question, "How far away is this number from zero?" is summed up simply with two bars placed around the number.

$$|5| = 5$$

$$|-5| = 5$$

That's the story behind an absolute value, but it is easy to just think of it like this. The power of the absolute value bars is that it makes the number inside a positive. Simplify the following absolute values. I've done the first one for you.

$$|-7| = 7$$

l) $|17| =$

m) $|-17| =$

n) $|-3| =$

o) $|-52| =$

p) $|258| =$

The idea can be carried over to variables. For instance:

Evaluate $|x|$ when $x = -3$. So: $|-3| = 3$.

Work a couple on your own.

Evaluate $|y|$ when:

q) $y = 9$

r) $y = -9$

On this next problem, the negative sign is on the outside, so it won't be removed.

Evaluate $-|v|$ when:

s) $v = 8$

t) $v = -8$

Here's the last idea for this activity. Absolute values are a grouping symbol. Therefore, when working the order of operations, absolute value symbols come first. We know that absolute value disregard the negative sign. But, be careful. We don't drop the negative until any work inside the absolute value has been done.

Look at the following problem.

$$|-12 + 2(3 - 2)|$$

We don't drop the negatives until everything has been cleaned up.

$$|-12 + 2(3 - 2)| = |-12 + 2(1)|$$

$$|-12 + 2|$$

$$|-10|$$

Now that we've cleaned up, we use the power of the absolute value sign.

$$|-10| = 10$$

Try these. Simplify.

u) $|-2(3 + 5) - 15 \div 3|$

v) $9 - |15 - 3(4 - 2)|$

w) $3(6) - 5^2 + |7(3 - 8) + 14 \div 7| + 11$

Algebra I

Active Lesson: 1.3b

Adding and subtracting negative numbers can give people a lot of difficulty. There are various ways to try and teach the idea. As I mentioned before, I think of it like money. A negative sign means debt.

$$-10 + 8$$

You owe 10 dollars and then you get 8. Your debt is now -2.

$$-10 + 8 = -2$$

Try these:

a) $-7 + 5 =$

b) $-21 + 7 =$

In these next problems, you will get out of debt:

c) $-5 + 8 =$

d) $-21 + 30 =$

e) $-1 + 15 =$

Look at this problem:

$$-10 - 8$$

You owe 10 dollars and then you owe 8 more. Your debt is now -18.

$$-10 - 8 = -18$$

Work these:

f) $-11 - 9 =$

g) $-8 - 3 =$

h) $-5 - 7 =$

Next, I've mixed them up. Work these. Remember, think of it as debt.

i) $-19 - 2 =$

j) $-1 + 18 =$

k) $14 - 7 =$

l) $-16 + 8 =$

m) $-5 - 6 =$

Secondly, we never want there to be more than one addition or subtraction sign in a row. So, we'll begin a problem by fixing that. The key is to think of negative signs as opposites. Addition signs don't change anything. Therefore:

$- +$:The opposite of a positive is: $-$

$+ -$:The positive doesn't change the negative so: $-$

$- -$:The opposite of an opposite: $+$

$+ +$:Two positives don't change anything: $+$

That's the logic, but here's the summary:

$- +$ will be $-$

$+ -$ will be $-$

$- -$ will be $+$

$+ +$ will be $+$

Simply the following. Turn the two signs into one.

n) $\quad -+5 =$

o) $\quad --14 =$

p) $\quad +-2 =$

q) $\quad -+9 =$

r) $\quad ++6 =$

In these next problems, the parentheses don't change anything. So, ignore the parentheses and simplify the two signs in a row.

s) $\quad -(+8) =$

t) $\quad +(+10) =$

u) $\quad -(-12) =$

v) $\quad +(-1) =$

Now, simplify the following by removing the double signs. For instance:

$$8 + (-6) = 8 - 6 = 2$$

w) $\quad 17 - (-2) =$

x) $\quad 10 - (+7) =$

y) $\quad 6 + (+5) =$

z) $-6 - (-7) =$

aa) $-41 + (-3) =$

bb) $-20 - (+9) =$

cc) $17 + (-11) =$

Finally, follow the order of operations to simplify these. Whenever you get two signs in a row, use the ideas that we've examined above.

$$9 + (-5 - 1) = 9 + (-6)$$
$$9 - 6 = 3$$

Work these:

dd) $22 + (-11 - 3) =$

ee) $-5 + (-19 - 2) =$

ff) $17 - (3 + 13) =$

gg) $-2 - (-12 + 6) =$

hh) $-40 - (-25 - 5) =$

ii) $20 - (-12 + 8) - 3 =$

Algebra I

Active Lesson: 1.4a

Multiply the following:

a) $2 \cdot 5 =$

b) $5 \cdot 2 =$

c) $1 \cdot 7 =$

d) $7 \cdot 1 =$

e) $3 \cdot 2 \cdot 1 =$

f) $1 \cdot 3 \cdot 2 =$

g) $2 \cdot 3 \cdot 1 =$

h)   Does the order in which we multiply matter?

We looked at adding and subtracting positive and negative numbers. Now we want to look at multiplying and dividing them. There is a pattern that can be memorized, but let's examine where it comes from.

Let's multiply:

$$-5 \cdot -3$$

The signs can be thought of like numbers, and since the order we multiply numbers doesn't matter, we can move the negatives to the front.

$$--5 \cdot 3 =$$

If we multiply the numbers then we get 15. And we learned in the last lesson that two negatives in a row is a positive. So:

$$-- 5 \cdot 3 = +15$$

For these problems, show how to move the negatives to the front. You won't normally have to, but I want you to get comfortable with where the final sign comes from.

i) $-3 \cdot -8 =$

j) $-9 \cdot -7 =$

k) $-6 \cdot -11 =$

l)   What will the final sign always be if we multiply two negative numbers together?

Next, let's multiply positive and negative numbers.

$$-6 \cdot 3$$

Move the signs to the front.

$$- + 6 \cdot 3$$

The numbers make 18. We learned previously that a negative multiplied by a positive gives a negative.

$$- + 6 \cdot 3 = -18$$

Try some. Again, show the signs moving to the front.

$-11 \cdot 5 =$

m) $6 \cdot -7 =$

n) $-8 \cdot 3 =$

o) $2 \cdot -4 =$

p)   What will the final sign always be if we multiply a positive number by a negative number?

The same ideas carry over to division. Move the signs to the front.

$$\frac{-15}{5}$$

$$-+\frac{15}{5}$$

15 divided by 5 is 3. A negative times a positive makes a negative. So:

$$-+\frac{15}{5} = -3$$

Divide the following:

q) $\dfrac{-27}{3} =$

r) $\dfrac{45}{-9} =$

s) $\dfrac{-42}{-7} =$

t) $-55 \div 11 =$

u) $-81 \div -9 =$

Next, let's look at exponents.

$$(-2)^4$$

This is really the same as:

$$-2 \cdot -2 \cdot -2 \cdot -2$$

Let's move the negatives out front.

$$----2 \cdot 2 \cdot 2 \cdot 2$$

Two negatives in a row make a positive.

$$\underset{\bigcirc\bigcirc}{- - - -}2 \cdot 2 \cdot 2 \cdot 2$$

So:

$$++2 \cdot 2 \cdot 2 \cdot 2 = 16$$

v) Try one. Move the negatives out front and then simplify.

$$(-2)^6$$

w) What will the final sign be any time you have an even exponent?

Let's work an odd exponent.

$$(-2)^3$$
$$-2 \cdot -2 \cdot -2 =$$

I've moved the negatives out front.

$$\underset{\bigcirc}{- -}-2 \cdot 2 \cdot 2 =$$

The first two make a positive.

$$\underset{\bigcirc}{+ -}2 \cdot 2 \cdot 2 =$$

A positive and a negative make a negative.

$$-2 \cdot 2 \cdot 2 = -8$$

x) Try one. Move the exponents out front.

$$(-2)^5 =$$

y) What will the final sign be any time you have an odd exponent?

Work these problems:

z) $(-3)^3 =$

aa) $(-5)^2 =$

bb) $(-4)^4 =$

cc) $(-6)^3 =$

Next, we will simplify expressions involving negative numbers. Remember to follow the order of operations.

dd) $-2(5-8) + (-3)^2$

ee) $20 \div 2 - 3(10-7) + 4(2-6) + (-2)^3$

Finally, we will evaluate expressions. This time they will involve negative numbers. Remember to substitute the number inside of a parenthesis.

ff)   Evaluate $x + 4$ when $x = -7$.

gg)   Evaluate $-(v - 8)$ when $v = -3$.

hh)   Evaluate $x^2 - 3x + 15$ when $x = -2$.

ii)   Evaluate $-2x^2 + 6x - 10$ when $x = -1$.

These last two problems have two variables.

jj)   Evaluate $(x - y)^2$ when $x = 3$ and $y = -2$.

kk)   Evaluate $(x + y)^3$ when $x = -2$ and $y = -3$.

Algebra I

Active Lesson: 1.4b

As I've mentioned, math is a language. And just as we could translate between English and Spanish or English and Japanese, we can translate between English and Math.

Translate the following into math language:

a)   The sum of 17 and -2.

b)   The product of 6 times 15.

c)   The quotient of 36 and -6.

d)   5 subtracted from 27.

e)   9 less than 12.

The translation typically works moving word by word from left to right. But we've previously seen that there are exceptions. "Subtracted from" and "less than" reverses the numbers. However, "difference of" is subtraction which doesn't reverse. It goes left to right. Translate the following:

f)   The difference of 80 and 45.

g)   The difference of 64 and 32.

A more difficult translation is when two different operations are performed. First, one operation must be completed and then the second. To indicate what must done first, we use grouping symbols. Notice this example.

The sum of 7 and -9, increased by 4.

$$[7 + (-9)] + 4$$

Translate the following on your own.

h)   The sum of 13 and -17, increased by 8.

i)   The sum of -4 and -7, increased by 2.

j)   The difference of 12 and 3, increased by 11.

k)   The sum of 19 and -2, decreased by 10.

Next, translate some which involve multiplication. (The multiplication must be done first.)

l)   The product of 3 and 2, increased by 5.

m)   The product of -4 and 8, decreased by 12.

And here are a couple which involve division. Translate the following:

n)   The quotient of 42 and 7, increased by 5.

o)   The quotient of 72 and -9, decreased by 3.

Finally, we will end with some basic word problems. Translate the problems. You don't need to solve them.

p)   The outdoor temperature was 70 degrees in the morning. It rose by 15 degrees at lunch time. What was the temperature?

q)   During a weekend hiking trip, two groups started at the same point and followed the same trail. One group travelled 22 miles. Another group travelled 14. What was the difference in how far they hiked?

r)   On a cold morning, the temperature was -2 degrees. Later, the afternoon temperature dropped to -8. What was the difference between the morning and afternoon temperatures?

s)   It costs $5 per student to go to a play. 10 students from an English class are going to attend. How much money will it cost them? (This is multiplication.)

t)   Another class spent $65 on tickets to the play. If tickets were $5 each, how many students attended? (This is division.)

In this activity, we are going to refresh your skills with fractions. It seems if no one enjoys working with fractions, but the concepts here are very important as we move into algebra. All of the ideas will reappear frequently, and the better that you understand them here, the easier the algebraic version will be.

<u>Equivalent Fractions</u>

Fractions are like slices of a pie. Here is the fraction $\frac{1}{3}$.

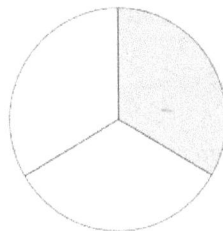

The bottom of the fraction is called the denominator, and it tells us how we have sliced the pie. With $\frac{1}{3}$, we have sliced the pie into three pieces. The top of the fraction is called the numerator, and it tells us how many slices we have. With $\frac{1}{3}$, we have one of the three slices.

However, we could slice the same pie differently. This is $\frac{2}{6}$.

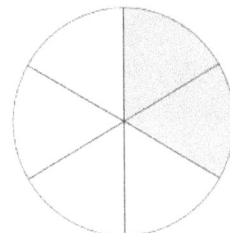

Although there are now six slices of pie, notice that our portion is still the same.

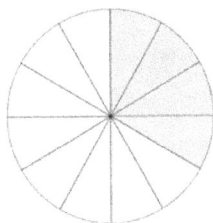

a)   I've cut the pie into another equivalent fraction. What is the fraction?

In each case, the fractions are the same.

$$\frac{1}{3} = \frac{2}{6}$$

But, how do we find equivalent fractions. These images tell us.

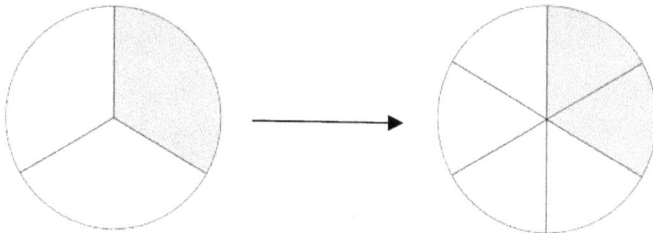

We doubled the number of slices. And that doubled the amount which we had.

$$\frac{1 \cdot 2}{3 \cdot 2} = \frac{2}{6}$$

It is like an equation. If we double the bottom, we must double the top. And there are any number of
b)   possible equivalent fractions. We could have tripled the slices. Finish the equivalent fraction:

$$\frac{1 \cdot 3}{3 \cdot 3} =$$

c)   If we start with $\frac{2}{5}$, which of the following would be an equivalent fraction?

   a) $\frac{8}{15}$

   b) $\frac{6}{15}$

   c) $\frac{6}{20}$

   d) $\frac{12}{20}$

d)   Give three equivalent fractions for $\frac{2}{3}$?

We can, of course, go the other direction.

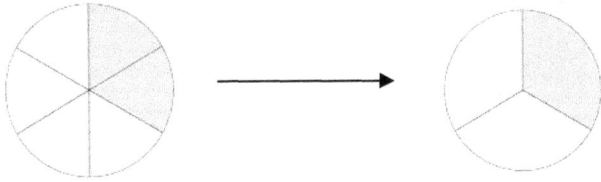

Here, we have cut half as many slices, so we have half the number of pieces. But to do this with math requires fractions. Turn the numerator and denominator into their prime factorization.

$$\frac{2}{6} = \frac{2}{2 \cdot 3}$$

If you have any matching factors in the numerator and denominator, cancel them.

$$\frac{2}{6} = \frac{\cancel{2}}{\cancel{2} \cdot 3} = \frac{1}{3}$$

Notice that we don't cancel the numerator down to nothing. Reducing leaves a 1 behind.

e) Select the correct reduced fraction. $\frac{15}{30} = \frac{3 \cdot 5}{2 \cdot 3 \cdot 5}$

   a) $\frac{3}{10}$
   b) $\frac{1}{15}$
   c) $\frac{1}{3}$
   d) $\frac{1}{2}$

A prime factorization tree is a helpful tool for fractions. Slice a number down to its prime numbers.

For even numbers, I keep dividing by 2. If two doesn't work, I try 3. Or if it ends in zero or five, I divide by 5. There are, of course, other possibilities, but these will typically help you get started. Notice on the right that the 2 was prime but 15 wasn't. So, I divided the 15 again.

Reduce the following fractions. I've given you the factor tree for the first two problems.

$$\frac{21}{49}$$

f)

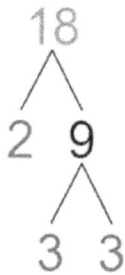 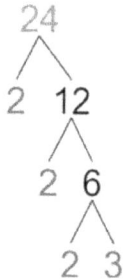

After you reduce, you'll have two factors left in the denominator. Simply multiply those factors back together.

$$\frac{18}{24}$$

g)

h)

$$\frac{40}{48}$$

This next problem has a negative out front. Just work the fraction like normal and return the negative to the final fraction.

$$-\frac{45}{54}$$

i)

Try one more. This time the problem involves large numbers. However, the process remains the same. Use your tree to create the prime factorization for the numerator and denominator and cancel any matches.

j)
$$-\frac{96}{144}$$

We can also extend the idea to algebra.

$$\frac{7x}{7y}$$

The x and the y are just factors.

$$\frac{7 \cdot x}{7 \cdot y}$$

We cancel any matching factors and get:

$$\frac{7 \cdot x}{7 \cdot y} = \frac{x}{y}$$

Try a couple.

k) $\dfrac{2a}{2b} =$

l) $\dfrac{3x}{6y} =$

Multiplying Fractions

Multiply the following:

m) $8 \cdot \dfrac{1}{2} =$

n) $20 \cdot \dfrac{1}{2} =$

Multiplying by half means that we end with half of what we started with. Suppose we started with a fraction like $\dfrac{1}{4}$.

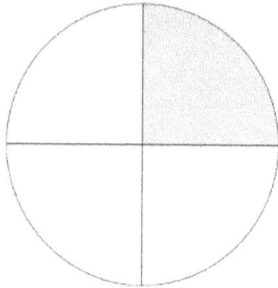

Multiplying by $\dfrac{1}{2}$ should give us half as much as when we started.

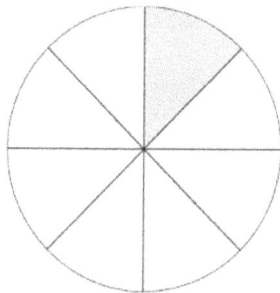

So, how does the math work.

$$\frac{1}{4} \cdot \frac{1}{2} =$$

We simply multiply across the top, and we multiply across the bottom.

$$\frac{1}{4} \cdot \frac{1}{2} = \frac{1}{8}$$

(Notice that this matches our expected slice of pie.)

Multiply the following fractions.

o) $\dfrac{1}{3} \cdot \dfrac{1}{2} =$

We can multiply by any fraction and it follows the same concept.

p) $\dfrac{1}{4}\cdot\dfrac{1}{5}=$

If you multiply across and your answer can be reduced, follow the procedure for simplifying fractions which we saw earlier.

q) $\dfrac{2}{3}\cdot\dfrac{1}{6}=$

r) $\dfrac{3}{10}\cdot\dfrac{4}{5}=$

There is a negative in this next problem. Just bring the negative out front.

s) $-\dfrac{6}{7}\cdot\dfrac{3}{8}=$

This problem has two negatives. Remember, if you multiply two negative numbers you get a positive.

t) $-\dfrac{5}{6}\cdot-\dfrac{1}{5}=$

Again, this can be extended to algebra. Just multiply across the tops and across the bottoms.

$$\dfrac{2}{3}\cdot\dfrac{x}{5}=\dfrac{2x}{15}$$

Try a couple:

u) $\dfrac{1}{7} \cdot \dfrac{y}{6} =$

This next problem can reduce.

v) $\dfrac{x}{10} \cdot \dfrac{2}{5} =$

Finally, let's look at division of fractions. Suppose I want to work the following problem:

$$\frac{1}{4} \div 2$$

This would mean that I have one-fourth of a slice and I want to share it among two people.

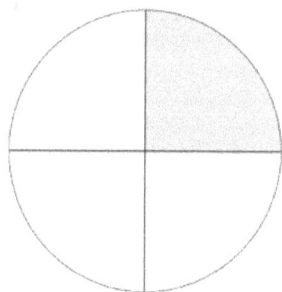

$\div 2 =$

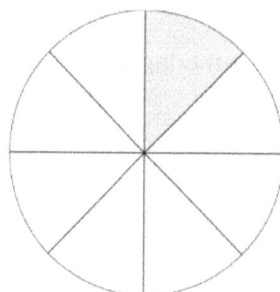

$$\frac{1}{4} \div 2 = \frac{1}{8}$$

Notice that this is the same answer we got when we multiplied by one half.

$$\frac{1}{4} \div 2 = \frac{1}{4} \cdot \frac{1}{2} = \frac{1}{8}$$

Dividing by fractions is hard to wrap our mind around. However, what we need to see is that we get the same answer if we multiply by the number turned upside down (the reciprocal). This always works and creates a strategy for dividing fractions: Keep, Change, Flip.

Keep the first number.

Change the sign to multiplication.

Flip the last number upside down.

Let me work one.

$$\frac{1}{4} \div \frac{2}{3} =$$

Keep the $\frac{1}{4}$.

Change to multiplication.

Flip the $\frac{2}{3}$ upside down and make it $\frac{3}{2}$.

$$\frac{1}{4} \cdot \frac{3}{2} = \frac{3}{8}$$

Try some.

w) $\quad \frac{1}{3} \div \frac{2}{5} =$

On your next problem, the final answer will reduce.

x) $\quad \frac{3}{4} \div \frac{3}{5} =$

This problem involves a negative number. Just bring the negative over to the final answer.

y) $\quad -\frac{4}{7} \div \frac{2}{5} =$

And this problem has two negatives. Remember, two negatives multiply to get a positive.

z) $\quad -\frac{7}{8} \div -\frac{7}{4} =$

Work one with more difficult fractions. After multiplying across the top and bottom, reduce down.

aa) $\dfrac{27}{40} \div \dfrac{72}{55} =$

And, finally, the ideas can again be extended to algebra.

$$\frac{x}{4} \div \frac{1}{3} = \frac{x}{4} \cdot \frac{3}{1} = \frac{3x}{4}$$

Try these:

bb) $\dfrac{1}{5} \div \dfrac{3}{7x} =$

cc) $\dfrac{5}{4} \div \dfrac{2x}{3} =$

dd) $\dfrac{y}{8} \div \dfrac{3}{12} =$

Algebra I

Active Lesson: 1.5b

Our first idea here is nothing new. A complex fraction is a fraction over a fraction.

$$\frac{\frac{2}{5}}{\frac{1}{3}} =$$

But a fraction bar is the same thing as division. So:

$$\frac{\frac{2}{5}}{\frac{1}{3}} = \frac{2}{5} \div \frac{1}{3}$$

And this is a problem we already know how to work. Keep. Change. Flip.

$$\frac{2}{5} \div \frac{1}{3} = \frac{2}{5} \cdot \frac{3}{1} = \frac{6}{5}$$

Work these:

a) $\dfrac{\frac{2}{7}}{\frac{3}{4}} =$

b) $\dfrac{\frac{4}{5}}{\frac{8}{9}} =$

As always, we can extend the idea to algebra.

$$\frac{\frac{x}{3}}{\frac{4}{5}} = \frac{x}{3} \div \frac{4}{5} = \frac{x}{3} \cdot \frac{5}{4} = \frac{5x}{12}$$

Simplify these:

c) $\dfrac{\frac{y}{3}}{\frac{2}{5}} =$

On this problem, the numbers simplify.

d) $\dfrac{\frac{2x}{7}}{\frac{6}{14}} =$

With this problem, remember that variables are just a factor and can be cancelled too.

e) $\dfrac{\frac{xy}{2}}{\frac{y}{3}} =$

f) $\dfrac{\frac{3y}{5}}{\frac{6xy}{15}} =$

Next, we want to simplify complex fractions which involve the order of operations. We know the fraction bar is division. But before we divide, simplify the numerator and the denominator.

$$\frac{2^2 - 3(4)}{6 + 2(3+2)} = \frac{4 - 12}{6 + 2(5)} = \frac{-8}{16}$$

I followed the order of operations to simplify both the top and the bottom. Now, I simply have a fraction which I will reduce.

$$\frac{-8}{16} = -\frac{2 \cdot 2 \cdot 2}{2 \cdot 2 \cdot 2 \cdot 2} = -\frac{1}{2}$$

Try these:

g) $\dfrac{8 - 5(3)}{7^2 - 7} =$

h) $\dfrac{6^2 + (-8)(-8)}{(-5)^2}$

On this problem, you will have a negative over a negative. Bring the negatives out front and they will cancel out.

i) $\dfrac{(-5)(6) + (-3)(8)}{2(8 - 10)}$

We will end by translating two-step quotient problems. To translate the quotient of 15 and 3, we would write:

$$\frac{15}{3}$$

j) Translate this: The quotient of x and y.

Now, let's make it a two-step problem. The quotient of the difference of x and z, and y. Notice that we are first taking a difference and then we are dividing.

$$\frac{x - z}{y}$$

Try these:

k) The quotient of the difference of x and 2, and y.

l) The quotient of the sum of x and y, and z.

Algebra I

Active Lesson: 1.6a

As we discussed before, fractions are like slices of pie. This is $\frac{1}{4}$.

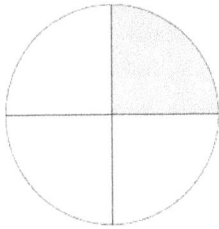

The bottom (denominator) is how the slices have been cut and the top (numerator) is how many slices we have. It wouldn't make sense to try to add slices that were different sizes. But if the slices are the same, you can.

$$\frac{1}{4} + \frac{2}{4}$$

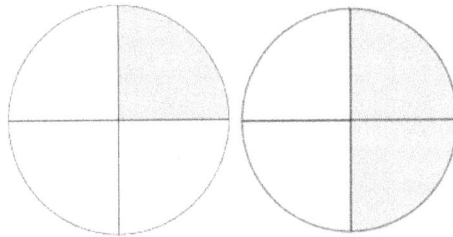

The size of the slices is the same. And, if we add them, we now have three. So:

$$\frac{1}{4} + \frac{2}{4} = \frac{3}{4}$$

Mathematically, it works like this. If the denominators are the same, we can simply add the numerators. Here's another.

$$\frac{2}{5} + \frac{1}{5} = \frac{3}{5}$$

Try some:

a) $\frac{3}{7} + \frac{2}{7} =$

b) $\frac{3}{12} + \frac{2}{12} =$

In this next problem, the new fraction can be reduced.

c) $\dfrac{1}{6} + \dfrac{2}{6} =$

Subtraction follows the exact same idea. If the denominators match, just subtract.

$$\frac{3}{5} - \frac{1}{5} = \frac{2}{5}$$

Try these:

d) $\dfrac{7}{11} - \dfrac{2}{11} =$

e) $\dfrac{15}{16} - \dfrac{2}{16} =$

On this problem, your answer can be reduced.

f) $\dfrac{8}{9} - \dfrac{2}{9} =$

This problem gets a negative for an answer. I know that is a strange idea when thinking of slices of pie, but mathematically it is okay.

g) $\dfrac{1}{4} - \dfrac{2}{4} =$

As always, we can extend the concepts to algebra.

$$\frac{x}{3} + \frac{1}{3}$$

It looks strange with an x, but it just means that I don't know how many slices I have in the first fraction. However, the denominators are the same, so I simply add.

$$\frac{x}{3} + \frac{1}{3} = \frac{x+1}{3}$$

But because the x and the 1 aren't like terms, this is the best I can do. Try some:

h) $\dfrac{y}{5} + \dfrac{3}{5} =$

i) $\dfrac{7}{15} + \dfrac{r}{15} =$

Subtraction works the same.

j) $\dfrac{x}{3} - \dfrac{1}{3} =$

k) $\dfrac{z}{7} - \dfrac{2}{7} =$

It is easy to add or subtract fractions when the denominators are the same. But what do we do when the denominators aren't the same?

$$\frac{1}{4} + \frac{1}{3}$$

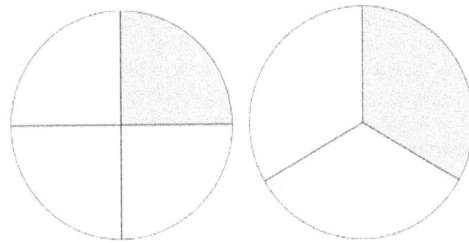

Adding two different sizes of pie would be meaningless. Instead, we need to re-slice the pieces so they match. In other words, we need to get a common denominator.

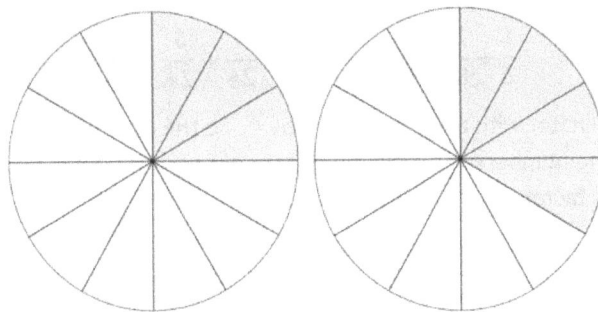

Now, with the slices the same, I can simply add.

$$\frac{3}{12} + \frac{4}{12} = \frac{7}{12}$$

But how did I determine the correct way to cut the pies? I used the same concept which we saw when we found Least Common Multiples. I found the lowest number that both denominators could multiply into. (Here it will be called finding the Least Common Denominator.) Here's a reminder of how we found the number.

Here are the prime factorizations of each of the original denominators.

$4: 2 \cdot 2$

$3: 3$

The first number they multiply into is the product of all their factors.

$$2 \cdot 2 \cdot 3 = 12$$

Multiplying all the factors would always give a common denominator, but not necessarily the least common denominator. Here is how we can be sure we find the LCD.

$$\frac{1}{6} + \frac{1}{8} =$$

Here are the prime factorizations.

$6: 2 \cdot 3$

$8: 2 \cdot 2 \cdot 2$

And, if they share any factors, circle them.

$6: \boxed{2} \cdot 3$
$8: \boxed{2} \cdot 2 \cdot 2$

Then, when you multiply your factors, any numbers in a circle only get counted once. So:

$2 \cdot 3 \cdot 2 \cdot 2 = 24$

Twenty-four is the lowest denominator that will work. It is the LCD- Least Common Denominator. Remember, you must make each fraction into an equivalent fraction. In other words, you can't just make the denominators 24 without changing the numerators.

$$\frac{1}{6} + \frac{1}{8} = \frac{1 \cdot 4}{6 \cdot 4} + \frac{1 \cdot 3}{8 \cdot 3} = \frac{4}{24} + \frac{3}{24} = \frac{7}{24}$$

Notice that for the denominator with a 6, I multiplied by 4. To the denominator with an 8, I multiplied by 3. Those are, of course, the numbers I need to multiply to get each to equal 24, but there is another reason. Look at the prime factorizations.

$6: \boxed{2} \cdot 3$
$8: \boxed{2} \cdot 2 \cdot 2$

l)   Why did the 6 need to be multiplied by 4 and the 8 need to be multiplied by 3?

Try some. (You can find the LCD in your head if you see it. However, it is very useful to learn how to do the prime factorization method. It will help us with harder problems as we move forward.)

m) $\dfrac{1}{6} + \dfrac{1}{14} =$

n) $\dfrac{3}{10} + \dfrac{4}{5} =$

o) $\dfrac{7}{8} + \dfrac{3}{10} =$

Subtraction works the same:

p) $\dfrac{1}{6} - \dfrac{1}{14} =$

q) $\dfrac{5}{12} - \dfrac{1}{8} =$

On this problem, you will end with a negative fraction. That is okay.

r) $\dfrac{1}{15} - \dfrac{3}{5} =$

Again, we can extend this to algebra.

$$\frac{x}{4} + \frac{1}{3}$$

The LCD is 12. And, even though it is difficult to wrap your head around, the process will be exactly the same.

$$\frac{x \cdot 3}{4 \cdot 3} + \frac{1 \cdot 4}{3 \cdot 4}$$

$$\frac{3x}{12} + \frac{4}{12} = \frac{3x + 4}{12}$$

I know the answer looks strange, but we followed all the same rules as adding normal fractions. (The $3x$ and the 4 are not like terms and so they couldn't be combined.)

Try some.

s) $\dfrac{y}{6} + \dfrac{1}{14} =$

t) $\dfrac{3}{10} + \dfrac{x}{5} =$

u) $\dfrac{z}{8} + \dfrac{3}{10} =$

Subtraction works the same:

v) $\dfrac{x}{6} - \dfrac{1}{14} =$

w) $\dfrac{5y}{12} - \dfrac{1}{8} =$

x) $\dfrac{1}{15} - \dfrac{3z}{5} =$

Algebra I

Active Lesson: 1.6b

Simplify the following complex fraction. Start by simplify the numerator and the denominator as if they were two different problems. Then, you are dividing a fraction, so Keep, Change, Flip.

a) $\dfrac{\left(\frac{1}{3}\right)^2}{6-2^2}$

Work the following fraction problems:

b) $\dfrac{1}{2}+\dfrac{1}{3}=$

c) $\dfrac{1}{3}+\dfrac{1}{4}=$

d) $\dfrac{5}{6}\div\dfrac{7}{12}=$

Here is another version of a complex fraction.

$$\frac{\frac{1}{2}+\frac{1}{3}}{\frac{1}{3}+\frac{1}{4}}$$

It looks terrible, but we already know how to do it. Simply, work it in three parts. First, the top. Second, the bottom. Third, divide your first answer by your second answer. (Keep, Change, Flip.) Notice that I already had you work the individual parts. Your answer to the third part is the answer to the problem.

Try some on your own.

$$\frac{\frac{2}{3}+\frac{1}{5}}{\frac{3}{5}+\frac{1}{5}}$$

First, work the top.

e)

Second, work the bottom.

f)

Third, take your answers and Keep, Change, Flip.

g)

Work another.

$$\frac{\frac{1}{2}-\frac{3}{7}}{\frac{3}{4}+\frac{4}{5}}$$

First, work the top.

h)

Second, work the bottom.

i)

Third, take your answers and Keep, Change, Flip.

j)

Next, we will evaluate expressions which include fractions. The idea is straightforward. As we do any time we evaluate, substitute the value inside of a parenthesis.

Evaluate $x + \frac{3}{4}$ when $x = \frac{1}{3}$.

k)
$$\left(\frac{1}{3}\right) + \frac{3}{4} =$$

If there is no multiplication to complete, just disregard the parenthesis. However, it is good practice to always substitute in using them.

Try a couple.

l)   Evaluate $y + \frac{3}{7}$ when $y = \frac{2}{3}$.

m)   Evaluate $x - \frac{7}{9}$ when $x = -\frac{1}{2}$.

Finally, we could be asked to evaluate an expression with fractions that includes two variables.

n) Evaluate x^2y where $x = \frac{1}{2}$ and $y = \frac{1}{3}$.

$$\left(\frac{1}{2}\right)^2 \left(\frac{1}{3}\right) =$$

Remember, order of operations would mean that we square $\frac{1}{2}$ first.

$$\left(\frac{1}{2}\right)^2 \left(\frac{1}{3}\right) = \frac{1}{4}\left(\frac{1}{3}\right) = \frac{1}{12}$$

Work these problems.

o) Evaluate x^2y where $x = \frac{2}{3}$ and $y = \frac{1}{4}$.

p) Evaluate x^2y where $x = -\frac{1}{2}$ and $y = -\frac{3}{7}$.

q) Evaluate x^2y^2 where $x = -\frac{1}{2}$ and $y = -\frac{3}{4}$.

Active Lesson: 1.7a

We are now going to begin a review of decimals. In actuality, decimals aren't used extensively in algebra, but they will appear on occasion.

Add the following fractions:

a)
$$\frac{1}{10} + \frac{1}{10} + \frac{1}{10} + \frac{1}{10} + \frac{1}{10} + \frac{1}{10} + \frac{1}{10} + \frac{1}{10} + \frac{1}{10} + \frac{1}{10} =$$

We've learned that the fraction bar in $\frac{1}{10}$ can also be thought of as division. Considering it that way, we have:

$$1 \div 10 = .1$$

And, .1 is called one tenth because ten of them would add to make 1. Likewise, $\frac{1}{100} = .01$ is called one hundredth because one hundred would add to make 1. Here is a table that includes the more common decimal places.

Hundred Thousand	Ten Thousand	Thousand	Hundred	Tens	Ones		Tenths	Hundredths	Thousandths	Ten-Thousandth	Hundred-Thousandth
						.					

As we did before, we can put numbers into their places. Here is the number 14.523

Hundred Thousand	Ten Thousand	Thousand	Hundred	Tens	Ones		Tenths	Hundredths	Thousandths	Ten-Thousandth	Hundred-Thousandth
				1	4	.	5	2	3		

Put the digits from the number 5782.1492 into their places on the chart:

b)

Hundred Thousand	Ten Thousand	Thousand	Hundred	Tens	Ones	Tenths	Hundredths	Thousandths	Ten-Thousandth	Hundred-Thousandth
						.				

Give the name of the place for each of the following digits:

c)   7

d)   4

e)   5

f)   9

When previously giving the names for numbers, we avoided the word "and." In mathematical naming, "and" is reserved for indicating the decimal point. So, the number 17.1 is:

Seventeen and one tenth.

Write the name of the following:

g) 125.9

h) 3.3

Giving a name to the decimal portion of the number is a bit tricky. First, find the place of the furthest decimal.

.14<u>5</u>

This is the thousandths place.

Second, name the decimal portion of the number as if there were no decimal.

145

One hundred forty-five.

Third, put the name of the number together with the name of the furthest decimal place.

One hundred forty-five thousandths.

Here's another example.

.56

First, the furthest decimal place is the hundredths.

Second, the number (without a decimal) would be fifty-six.

Third, put them together: fifty-six hundredths.

Try these:

i)   .16

j)   .412

k)   .07

If there is a number before the decimal point, we name that number as we've done before, we put *and*, and then name the decimal portion.

17.25

Seventeen *and* twenty-five hundredths.

145.023

One hundred forty-five *and* twenty-three thousandths.

Try these on your own:

l)   67.12

m)   475.18

n)   2045.003

o)   6.01

Next, we will take the concept in the other direction. Here is a number with a decimal written out:

One hundred seventeen and twenty-one thousandths.

And will be the decimal point. Left of the *and*, we name as we always would.

117.

To the right of the *and*, we write the number (which is underlined with one line), ending it at the decimal place (which is underlined with two lines).

twenty-one thousandths

117.021

Here's another:

Thirty-seven and <u>two</u> <u>hundredths</u>.

37.02

Try these:

p)   Four hundred twelve and fifteen thousandths.

q)   Five and twenty-five hundredths.

r)   Two thousand and six tenths.

s)   Six hundred ninety-nine and two thousandths.

Rounding numbers with decimals works exactly the same as we saw before. Let's walk through one:

Round 159.347 to the hundredths place.

First, find the place we want to round.

159.3<u>4</u>7

Look to the number behind it. If it is 5, 6, 7, 8, or 9, round up. If it is 0, 1, 2, 3, or 4, keep your number the same. Previously, we made any number which followed into zeros. Here, the zeros on the decimal aren't helpful so we just drop them.

$$159.35$$

Round 159.347 to the tenths place.

$$159.\underline{3}47$$

The number behind it is a 4. So:

$$159.3$$

Try some on your own:

t)   Round 631. 4492 to the hundredths place.

u)   Round 631. 4492 to the tenths place.

v)   Round 631. 4492 to the hundreds place. (Be careful on this one.)

w)   Round 631. 4892 to the nearest whole number. (This would mean the ones place.)

Finally, we will add and subtract decimals. This should be a review for you, so we will only look at it briefly.

$$24.456 + 12.37$$

First, line the numbers up by their decimal points.

```
    2 4 . 4 5 6
+   1 2 . 3 7 0
  ─────────────
```

Then, add each column, starting at the right and moving left.

```
          1
    2 4 . 4 5 6
+   1 2 . 3 7 0
  ─────────────
    3 6 . 8 2 6
```

In the second column, $5 + 7 = 12$. Whenever we go over 9, we need to add a carry. The two from the 12 goes below, and we carried the 1 to the next column to the left. It was then added with that column.

Try one. This problem will have two carries.

x) $36.193 + 15.46$

Subtraction follows the same process.

```
    2 4 . 4 5 6
−   1 2 . 3 7 0
  ─────────────
```

First line up the decimals from each number. Subtract the columns from right to left. (The top number minus the bottom number.) If top number is too small, borrow from the next column over.

```
          3 10
    2 4 . ⁄4 5 6
−   1 2 . 3 7 0
  ─────────────
              6
```

Here, $5 - 7$ in the second column wouldn't work. So, we borrow from the next column over. That makes it a 3 instead of a 4. That allows for us to add an extra 10 to our column.

```
              3  15
      2  4  .  4̸  5  6
  −   1  2  .  3  7  0
  ─────────────────────
      1  2  .  0  8  6
```

Try this one on your own.

y)   57.389 − 42.45

As we continue working with decimals, let's look at multiplying and dividing decimal numbers. First, multiplication.

The process of multiplying decimal numbers is exactly the same as any multiplication, with only one exception. We count up the number of decimal places for the numbers involved. Here's an example:

$$24.13$$
$$\times \quad 15.7$$

Multiply as if there were no decimals.

		2	4	1	3
×			1	5	7
+		1	6	8	9	1
+	1	2	0	6	5	
+	2	4	1	3		
=	3	7	8	8	4	1

The original problem had three decimals: one in 15.7, two in 24.13. Give three decimal places to your final answer.

$$378.841$$

Try a couple on your own:

a) $48.75 \cdot 13.4$

b) $186.841 \cdot 87.72$

Another important skill, and a much easier one, is multiplying decimals by multiples of 10.

$$186.841 \cdot 10 =$$

Because our number system is based on the number 10, multiplying by 10 simply moves the decimal place one spot to the right.

$$186.841 \cdot 10 = 1868.41$$

Multiplying by 100 moves it two spots to the right.

$$186.841 \cdot 100 = 18684.1$$

Work the following:

c) $4876.985 \cdot 10 =$

d) $4876.985 \cdot 100 =$

e) $4876.985 \cdot 1000 =$

Dividing number with decimals also follows the basic ideas of long division. The only exception is that we don't want the number we divide by (the divisor) to be a decimal.

$$24.456 \div 12.37$$

So, we remove the decimal by moving the decimal place. For 12.37, we move it two places. But we can't just change part of the problem. It turns out that if we move the decimal for the number we are dividing into (the dividend), everything will still divide the same as it would have before. Therefore, we make the dividend 2445.6.

```
            1.97704
  1237| 2445.60000
       1237
       12086
       11133
        9530
        8659
         8710
         8659
          510
            0
         5100
         4948
          152
```

Notice that the decimal is in a new location and it was brought directly up to the answer. (This long division didn't actually end; however, it was stopped at 5 decimal places.)

Divide these: (If the number doesn't stop, finish at three decimal places.)

f) $81.5 \div 10.5$

77

g) $247.685 \div 25.32$

Here you are dividing a negative number. A negative divided by a positive will give a negative number as the answer. (Ignore the negative when dividing and then add the negative to the solution.)

h) $-12.547 \div 2.3$

In this activity, we'll take one more look at decimals. We will begin with how to write decimals as fractions. To do so, we need to recall the names of the decimal places.

Tenths	Hundreths	Thousanths	Ten-Thousanth	Hundred-Thousanth

To write a decimal as a fraction, find the name of the last decimal place.

$$.54$$

This number ends in the hundredths place, so put the number 54 over 100.

$$\frac{54}{100}$$

If the fraction can be reduced, be sure to do so.

$$\frac{54}{100} = \frac{27}{50}$$

Here's another.

$$.758$$

The last decimal place is the thousandths, so:

$$\frac{758}{1000} = \frac{379}{500}$$

Write the following decimals as fractions.

a) .8   b) .25   c) .37   d) .1474

Next, we want to go in the other direction. To write a fraction as a decimal, we just need to remember that the fraction bar is a division bar.

$$\frac{5}{8} = 5 \div 8$$

To divide, use long division.

```
        0 . 6 2 5
    8  ) 5 . 0 0 0
         —
         0
         5   0
         —
         4   8
             2   0
             —
             1   6
                 4   0
                 —
                 4   0
                 ———
                 0
```

Convert the following fractions into a decimal.

e) $\dfrac{4}{5}$ f) $\dfrac{7}{8}$

Sometimes when we divide a fraction, we will find that the same numbers keep repeating:

$$\frac{1}{12} = .083333\ldots$$

Since the repeat would go forever, there's a shortcut. We put a line over the portion which repeats.

$$.08\overline{3}$$

The repeat can be more than a single number:

$$\frac{4}{11} = .363636\ldots$$

$$.\overline{36}$$

Even if the numerator is larger than the denominator, it can still occur.

$$\frac{27}{11} = 2.454545\ldots$$

$$2.\overline{45}$$

80

Write the following fractions as decimals. Indicate if they repeat.

g) $\dfrac{1}{3}$

h) $\dfrac{2}{11}$

i) $\dfrac{19}{8}$

Finally, we will convert percentages to decimals. A percentage is part of a whole. 100% is the whole. For instance:

$$15\% = \dfrac{15}{100}$$

And we know that a fraction can be written as a decimal.

$$100 \div 15 = .15$$

Because a percentage can always be written as a fraction over 100, we could always just divide the % by 100. This moves the decimal place two spots to the left.

$$15\% = .15$$

And:

$$27\% = .27$$

Write the following percentages as a decimal.

j) 68%

k) 16%

(When we think of this next one as part of a whole, it will seem odd. However, it still works the same.)

l)   115%

In a similar fashion, we could go the other direction, turning a decimal into a percentage. Just move the decimal place two spots to the right.

$$.35 = 35\%$$

Or:

$$.045 = 4.5\%$$

Change the following decimals to percents.

m) .12

n) .657

o) .1117

p) .003

Algebra I

Active Lesson: 1.8a

Mathematicians like patterns. As such, they noticed that numbers fall into a variety of categories. As you pass through algebra, understanding these number categories is very helpful. But before we look at the categories, we first want to have a quick refresher on square roots.

Here's the idea behind a square root. Mathematician wanted to answer this question: "What two of the same number multiply together to create this number?" Using math language, that entire question was summed up with the square root sign.

$$\sqrt{9}$$

What two of the same number multiply together to create the number 9? The answer if 3.

$$\sqrt{9} = 3$$

Answer these square roots:

a) $\sqrt{25}$

b) $\sqrt{36}$

c) $\sqrt{121}$

d) $\sqrt{169}$

As we will discuss later, square roots are not allowed to have negatives inside. However, a negative in front is fine. It just means you will find the answer to the square root and then add the negative sign.

$$-\sqrt{81} = -9$$

Answer these square roots:

e) $-\sqrt{49}$

f) $-\sqrt{64}$

Now we are ready to begin learning the different classifications of numbers.

Counting Numbers- these are the first numbers you learned as a child: 1, 2, 3, 4, 5.... Counting numbers exclude zero.

Whole Numbers- these are identical to the counting numbers except that they also include zero.

Integers- these include any number without a decimal or fraction.

$$...-3, -2, -1, 0, 1, 2, 3...$$

Our next classification is rational numbers.

Rational Numbers- these include any number which can be expressed as a fraction.

$$\frac{4}{5}, -\frac{17}{20}$$

Those are obvious, but these are rational numbers too:

$$5, .02$$

They make not look like it, but they can be expressed as fractions. Any integer can be a fraction simply by putting it over the number 1.

$$5 = \frac{5}{1}$$

And we previously learned how to make decimals into fractions.

$$.02 = \frac{2}{100}$$

Express the following numbers as fractions:

g) 121   h) .012   i) -6   j) .8

Our next classification is irrational numbers.

Irrational Numbers- These are numbers which cannot be expressed as a fraction. They cannot be expressed as a fraction because they contain decimals which never repeat.

$$3.1415926\ldots, -.171819\ldots$$

Rational vs. Irrational Numbers

We saw that rational numbers can be written as a fraction. In decimal form, rational numbers will always end or will have a repeated pattern.

Circle any of the following numbers which are rational. Put a square around any numbers which are irrational.

k) $4.\overline{125}$ 6.25 $2.7182818459\ldots$ $7.15423985407\ldots$

A key place where we find irrational numbers is square roots. If a number doesn't have a perfect square root it is an irrational number.

$\sqrt{21}$ is irrational. There are no two identical numbers which multiply to make 21. Using a calculator we find:

$\sqrt{21} = 4.582575695\ldots$

It makes a decimal which doesn't repeat.

l)   Circle any of the following numbers which are rational. (They will have a perfect square root.) Put a square around any numbers which are irrational.

$\sqrt{4}$

$\sqrt{48}$

$\sqrt{35}$

$\sqrt{225}$

We have one final classification (at least for this point in algebra).

Real Numbers- Any number which can be classified as a rational number or an irrational number.

The only exception to a real number which we will see in this course is a negative under a square root.

$$\sqrt{-9}$$

This doesn't exist and isn't a real number.

If you notice, our classification system keeps building outward.

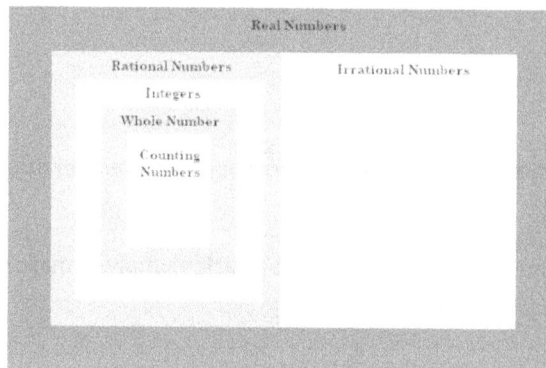

Counting numbers are also whole number, integers, rational numbers, and real numbers.

Whole numbers are also integers, rational numbers, and real numbers.

Irrational numbers are also real numbers.

For the numbers below, give the letter (or letters) which could classify the number. (Remember, it builds outward. A counting number would be everything except irrational.)

A-Counting Number   B-Whole Number   C-Integers   D-Rational Number

E-Irrational Number   F-Real Number

m) $\frac{7}{12}$

n) 5.148236427....

o) $12.\overline{3}$

p) -11

q) $\sqrt{15}$

r) $\sqrt{9}$

s) 7

t) 0

u) $\sqrt{-16}$

Algebra I

Active Lesson: 1.8b

Students will often have a difficult time identifying the location of fractions or decimals along a number line. In this activity, we will look at these ideas.

Decimals

To understand the location of decimals, we must remember that number lines move from left to right. The further a number is to the right then the larger the number is.

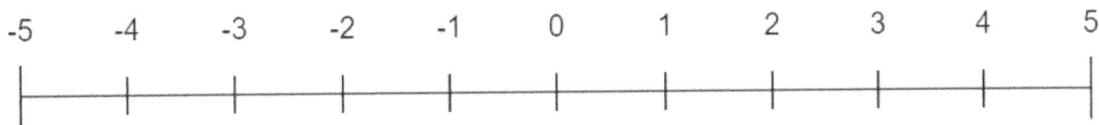

Although we haven't placed it precisely, 2.3 would be to the right of 2, and so it is larger. Circle the larger number:

a) 4.4 or 4

b) 3.1 or 3.7

c) 1 or .8

Next, use an inequality sign to indicate which is larger. (Remember, inequalities also read from left to right. < is less than. > is greater than.)

d) 5 ___ 5.7

e) 2.4 ___ 2

f) 1.6 ___ 0

Things get a little trickier with negatives. For instance, -2.7 is further left than -2. If it is further left on a number line, it is smaller. Circle the larger number:

g) -4.4 or -4

h) -3.1 or -3.7

i) -1 or -.8

Next, use an inequality sign to indicate which is larger.

$$j) \quad -5\_ -5.7$$

$$k) \quad -2.4\_ -2$$

$$l) \quad -1.6\_0$$

Let's look at the idea again. This time with further decimals.

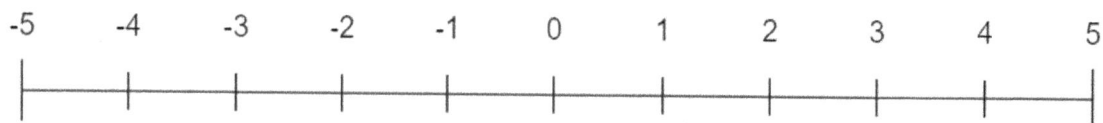

Adding a second (or third decimal) to a positive number means it has a little bit more.

$$1.21 > 1.2$$

$$4.55 < 4.551$$

And if it has a bit more, it is further to the right on a number line. Use an inequality sign to indicate which number is larger:

m) 5.33__5.3

n) 7.4__7.46

o) 1.2221__1.222

A negative number with a bit more would actually be further to the left on a number line. So:

$$-2.11 < -2.1$$

$$-5.75 < -5.72$$

Use an inequality sign to indicate which number is larger:

p) -5.33___ -5.3

q) -7.4___ -7.46

r) -1.2221___ -1.222

In class, to place decimals on a number line, I allow students to approximate. For instance, 1.3 is larger than 1 but not quite halfway to the number 2.

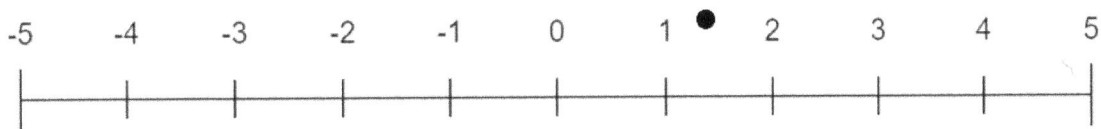

And, -3.55 is less than -3 and about half way to -4.

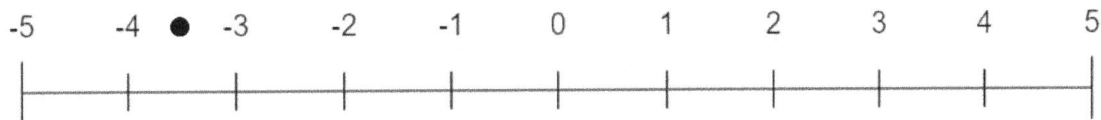

However, if you want to get more precise, remember that the first decimal place is the tenths. So, we could add 10 small lines between the numbers. Here is .2 marked.

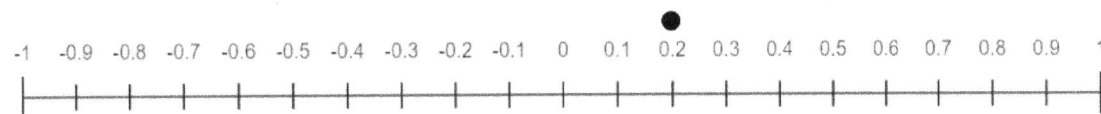

For a number with hundredths, actually marking the lines would get quite difficult. Again, I would approximate. Here is .55. A bit more than half way to .6.

Mark the following numbers on a number line. (Use the approximation method.)

s)   4.2

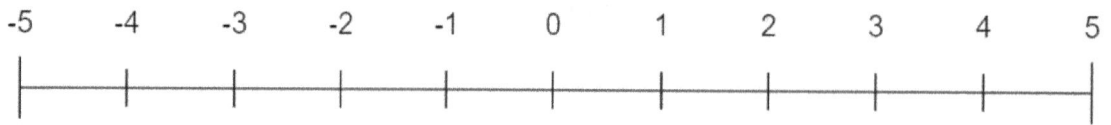

t)   4.5

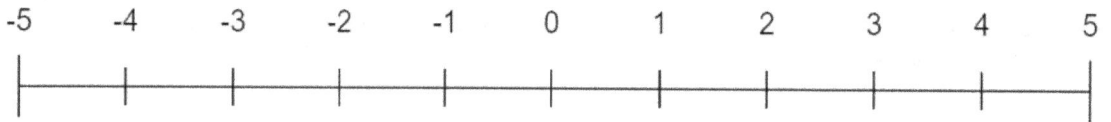

u)   -4.2

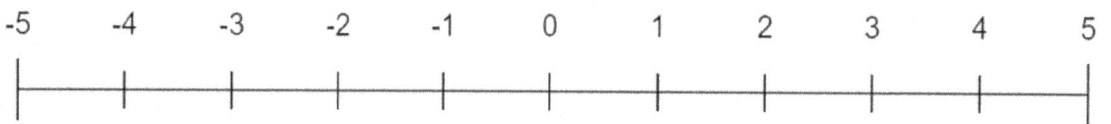

v)   -4.5

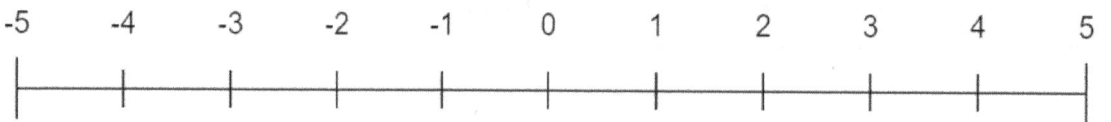

w)   .058

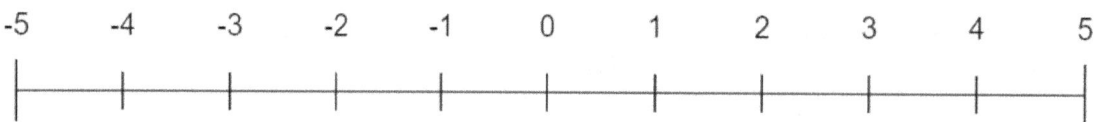

x)   -2.882

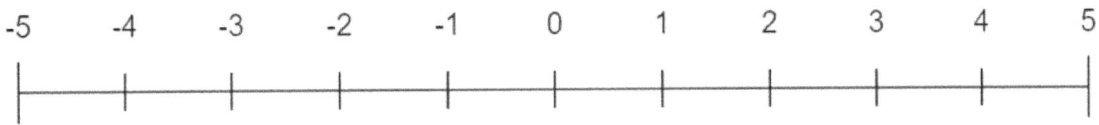

Fractions

To locate fractions, you may find it easy to visualize them as part of a whole, and mark it accordingly. For instance, $\frac{3}{10}$ would be about half way between 0 and .5.

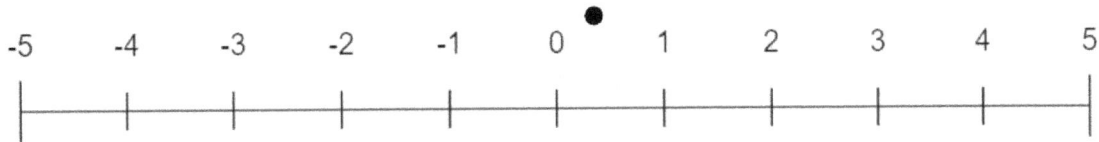

However, a much easier approach is to turn fractions into decimals and graph them on the number line just as we did before.

$$\frac{3}{10} = .3$$

Or:

$$\frac{13}{3} = 4.\overline{3}$$

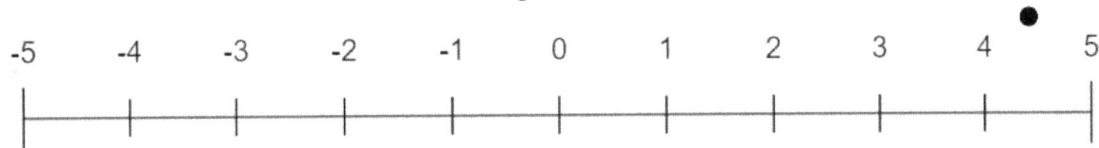

Or:

$$-\frac{12}{5} = -2.4$$

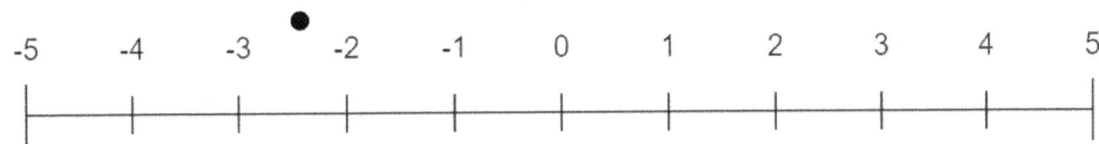

For the following, feel free to use your calculator to divide the fractions. Then, use the approximation approach to mark then on the number line.

y) $\dfrac{1}{5}$

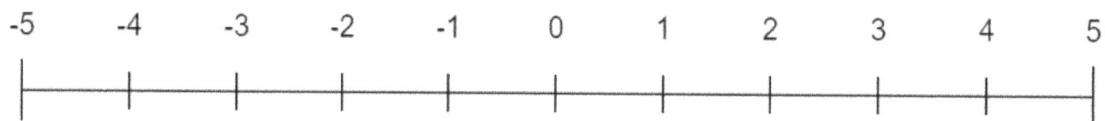

z) $\dfrac{8}{5}$

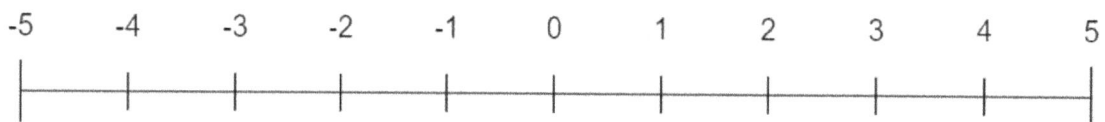

aa) $-\dfrac{9}{4}$

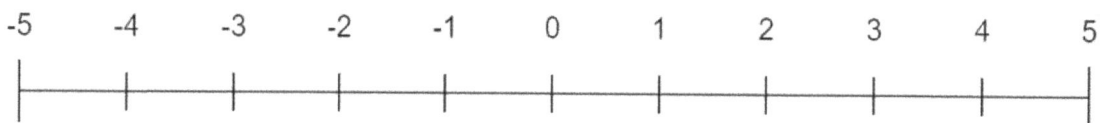

bb) $-\dfrac{10}{3}$

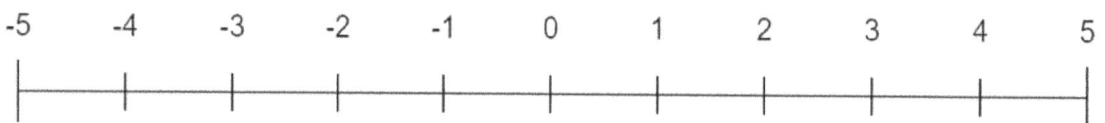

Algebra I

Active Lesson: 1.9a

When we do mathematical operations there are various ideas which always hold true. We call these truths—properties. In this activity, we will examine a number of important properties.

Commutative Property of Addition

Add the following:

a) $7 + 12 =$

b) $12 + 7 =$

c) $-2 + 9 =$

d) $9 + (-2) =$

e) Does the order in which we add two numbers matter?

A commute is a journey to work. If you drive from work to home or from home to work, you've travelled the same distance. That's how I remember the name of the commutative property.

The application to algebra involves combining like terms. We can add in any order.

$$8x + 3y + 2x - 5y$$
$$8x + 2x + 3y - 5y$$
$$10x - 2y$$

Try these:

f) $12v + 14u - 2v + 15u$

g) $-9r + 7s + 2r + 17s$

<u>Commutative Property of Multiplication</u>

Multiply the following:

h) $5 \cdot 6 =$

i) $6 \cdot 5 =$

j) $-3 \cdot 7 =$

k) $7 \cdot -3 =$

l)   Does the order in which we multiply two numbers matter?

This is the commutative property of multiplication. The application in algebra can be seen here.

$$x \cdot 4 = 16$$

We like to solve equations with the variable at the back. But using the commutative property of multiplication we can simply rearrange it.

$$4x = 16$$

<u>Associative Property of Addition</u>

Add the following numbers:

m) $4 + (2 + 3) =$

n) $(4 + 2) + 3 =$

o) $-6 + (7 + 3) =$

p) $(-6 + 7) + 3 =$

q)   Does it matter which group we add together first?

The parentheses are associating two number together. It doesn't matter who associates first. We get the same outcome.

The associative property of addition can save us time in algebra.

$$(-2x + 7x) + 2x$$

Because the order we associate doesn't matter. We could change the order to make the work easier.

$$7x + (-2x + 2x) = 7x$$

Add the following using the Associative Property of Addition.

r) $(-15y + 3y) + 15y$

s) $7v + (-7v + 8v)$

Associative Property of Multiplication

Multiply the following numbers.

t) $(4 \cdot 3) \cdot 2 =$

u) $4 \cdot (3 \cdot 2) =$

v) $(-3 \cdot 5) \cdot 6 =$

w) $-3 \cdot (5 \cdot 6) =$

x) Does it matter which group we multiply first?

In Algebra, the Associative Property of Multiplication can also make the work easier.

$$(\frac{1}{2} \cdot x) \cdot 2$$

$$\left(\frac{1}{2} \cdot 2\right) \cdot x = 1x$$

Identity Property of Addition

Add the following:

y) $8 + 0 =$

z) $-15 + 0 =$

aa) $1254 + 0 =$

bb)   What happens whenever you add zero to a number?

This is called the identity property of addition because when we add zero to a number that number keeps its original identity; it doesn't change.

The concept carries over to algebra. Try these:

cc) $5x + 0 =$

dd) $-17y + 0 =$

<u>Identity Property of Multiplication</u>

Multiply the following:

ee) $3 \cdot 1 =$

ff) $-7 \cdot 1 =$

gg) $57 \cdot 1 =$

hh) What happens when you multiply a number by 1?

This is called the Identity Property of Multiplication because when we multiply a number by one, the identity of that number doesn't change.

The application in algebra is the same. Multiply the following:

ii) $16x \cdot 1 =$

jj) $8y \cdot 1 =$

<u>Inverse Property of Addition</u>

The additive inverse of a number is its opposite. Add these opposites:

kk) $7 + (-7) =$

ll) $15 + (-15) =$

mm) $-9 + (9) =$

nn) What happens when you add a number to its additive inverse?

Again, we can use this property in Algebra. Simplify:

oo) $2x + (-2x) =$

pp) $-15y + 15y =$

Inverse Property of Multiplication

A multiplicative inverse is a bit different; it is a number flipped upside down. Complete the table below. I've done the first two.

Number	Multiplicative Inverse
7	$\frac{1}{7}$
-5	$-\frac{1}{5}$

qq) 12

rr) 9

ss) $\frac{1}{2}$

Multiply the following numbers by their multiplicative inverse.

tt) $6 \cdot \frac{1}{6} =$

uu) $15 \cdot \frac{1}{15} =$

vv) $-2 \cdot -\dfrac{1}{2} =$

ww) $\dfrac{1}{3} \cdot 3 =$

xx) What happens when we multiply a number by its multiplicative inverse?

In algebra, again, the concept remains the same.

$$5x \cdot \dfrac{1}{5x} = 1$$

Multiply the following:

yy) $8y \cdot \dfrac{1}{8y} =$

zz) $\dfrac{1}{7z} \cdot 7z =$

Multiplication by Zero

Multiply the following numbers by zero.

aaa) $8 \cdot 0 =$

bbb) $-3 \cdot 0 =$

ccc) $125 \cdot 0 =$

ddd) What happens whenever you multiply a number by zero?

The same is true in algebra.

Multiply the following:

eee) $18x \cdot 0 =$

fff) $(x - 4) \cdot 0 =$

Division of Zero

Division is taking an amount and splitting it into groups. To divide 15 by 3 means to take 15 and put it into 3 groups. Each group would get 5.

$$15 \div 3 = 5$$

What if we start with 0 and try to divide it into 3 groups.

$$0 \div 3$$

If there is nothing to share, everyone gets nothing.

$$0 \div 3 = 0$$

Divide the following:

ggg) $0 \div 8 =$

hhh) $0 \div 2 =$

This looks different, but it is still division.

iii) $\dfrac{0}{10} =$

These next two seems strange, but the result is the same.

jjj) $0 \div -5 =$

kkk) $\dfrac{0}{-5} =$

lll) What happens when we divide zero by a number?

Again, it remains the same in algebra.

$$0 \div 4x = 0$$

Simply the following.

mmm) $0 \div 12y =$

nnn) $\dfrac{0}{(2x - 5)} =$

Division by Zero

If division is taking a number and dividing it into groups, what happens if we divide by zero?

$$3 \div 0$$

This would mean that we start with 3 and divide it into 0 groups. But how can we split something among zero groups? It would mean that we have made the 3 disappear. So, it can't be done. Because of this, dividing by zero is called "Undefined." In other words, it is impossible.

Divide the following numbers by zero.

ooo) $5 \div 0$

ppp) $-25 \div 0$

qqq) $\dfrac{2}{0}$

rrr) $\dfrac{-8}{0}$

Lastly, the concept remains the same in algebra.

$$2x \div 0 =$$

Undefined.

Divide the following:

sss) $5r \div 0 =$

ttt) $\dfrac{(3y + 5)}{0} =$

Multiply the following:

a) $3 \cdot 5 =$

b) $3(2 + 3) =$

c) $(3 \cdot 2) + (3 \cdot 3) =$

It turns out that the following is true:

$$3(2 + 3) = (3 \cdot 2) + (3 \cdot 3)$$

Although it seems unnecessary, would could break a number into pieces and then multiply each piece.

$$3(2 + 3)$$

Here, I've multiplied the 3 to the first number inside and then to the second number inside. This is called the distributive property. It doesn't seem to have much of a point when multiplying two numbers, but in algebra it is very important.

$$5(x + 3)$$

$(x + 3)$ is a number, but I don't know what it is. The problem is the x. Here's where the distributive property comes in.

$$5(x + 3) = 5x + 15$$

Use the distributive property to simplify the following expressions:

d) $2(x + 5)$

e) $10(x - 2)$

f) $-3(x+4)$

g) $x(2+y)$

In these next two examples, we are actually multiplying a -1 through.

h) $-(3x+4)$

i) $-(2y-15)$

In these next problems, use the distributive property and then "clean-up" by combining any like terms.

j) $15+4(x-2)$

k) $-6-2(3x+7)$

l) $5x+3(6x-6)$

m) $-6y-6(2y+1)$

n) $3(2x-2)-6(5x+3)$

o) $12(2y+1)-(20y-8)$

Algebra I

Warm Up: 1.10a

I'm going to multiply the following fractions:

$$\frac{2}{3} \cdot \frac{1}{2}$$

I could multiply across the top and bottom and then reduce, but it is equally fine to reduce the 2's from the start:

$$\frac{\cancel{2}}{3} \cdot \frac{1}{\cancel{2}} = \frac{1}{3}$$

Reduce the following fractions by reducing diagonally:

a) $\frac{3}{5} \cdot \frac{2}{3} =$

b) $\frac{5}{7} \cdot \frac{2}{5} =$

The idea of reducing diagonally is a helpful idea in understanding our next topic. We want to convert between different units.

A conversion between two units can be set up as a fraction. For instance, 1 foot = 12 inches can be written as:

$$\frac{1\ foot}{12\ inches} \quad or \quad \frac{12\ inches}{1\ foot}$$

c)   1 yard has 3 feet. Write the conversion as two fractions.

To convert between units, we will set up a fraction problem. And we want to use the conversion as a fraction with one important condition. The unit we are trying to get rid of needs to be on the bottom of the fraction.

Convert 36 inches into feet.

$$36 \; inches \cdot \frac{1 \; foot}{12 \; inches}$$

I want to get rid of inches, so it is at the bottom of the conversion fraction. Next, the name of the units works just like a factor. They can cancel diagonally.

$$36 \; in\cancel{ches} \cdot \frac{1 \; foot}{12 \; in\cancel{ches}}$$

Now, multiply across and keep the name of the unit which hasn't been cancelled.

$$36 \; in\cancel{ches} \cdot \frac{1 \; foot}{12 \; in\cancel{ches}} = \frac{36 \; feet}{12}$$

Then reduce the numbers like normal.

$$\frac{36 \; feet}{12} = 3 \; feet$$

If you wanted to convert feet to inches, just use the conversion ratio upside down.

Convert 5 feet to inches.

$$5 \; f\cancel{eet} \cdot \frac{12 \; inches}{1 \; f\cancel{oot}} = \frac{60 \; inches}{1} = 60 \; inches$$

Use this method to complete the following conversions:

d)   Convert 72 inches to feet.

e)   Convert 9 feet to inches.

This same idea can be used to make any conversion. Here is a list of unit conversions in the U.S. system of measurement.

Length	1 foot (ft) = 12 inches (in) 1 yard (yd) = 3 feet (ft) 1 mile (mi) = 5280 feet (ft)	Weight	1 pound (lb) = 16 ounces (oz) 1 ton = 2000 pounds (lb)
Volume	3 teaspoons (t) = 1 tablespoon (T) 16 tablespoons (T) = 1 cup (C) 1 cup (C) = 8 fluid Ounces (fl. oz.) 1 pint (pt) = 2 cups (C) 1 quart (qt) = 2 pints (pt) 1 gallon (gal) = 4 quarts (qt)	Time	1 minute (min) = 60 seconds (sec) 1 hour (hr) = 60 minutes (min) 1 day = 24 hours (hr) 1 week (wk) = 7 days 1 year (yr) = 365 days

Convert 17 fluid ounces to cups. Here, I've divided the fraction to a decimal.

$$17 \; fl.oz. \cdot \frac{1 \; cup}{8 \; fl.oz.} = \frac{17 \; cup}{8} = 2.125 \; cups$$

Use this method and the chart above to make the following conversions:

f)   Convert 3150 pounds to tons.

g)   Convert 5 miles into feet.

The concept can be repeated to make multiple conversions.

Convert 3 days into minutes.

$$3 \; days \cdot \frac{24 \; hours}{1 \; day} \cdot \frac{60 \; minutes}{1 \; hour} = \frac{4320 \; minutes}{1} = 4320 \; minutes$$

Notice how the units continued to cancel until only minutes remained.

Try some on your own.

h)   Convert 5 miles into inches.

i)   Convert 2000 seconds into hours. (This will be a very small number.)

j)   Convert 4 pints into tablespoons.

The same concept can be used in the metric system. Here is a list of conversions.

Length		Mass		Capacity	
	1 kilometer (km) = 1000 m		1 kilogram (kg) = 1000 g		1 kiloliter (kL) = 1000 L
	1 hectometer (hm) = 100 m		1 hectogram (hg) = 100 g		1 hectoliter (hL) = 100 L
	1 dekameter (dam) = 10 m		1 dekagram (dag) = 10 g		1 dekaliter (daL) = 10 L
	1 meter = 1 m		1 gram (g) = 1 g		1 liter (L) = 1 L
	1 decimeter (dm) = .1 m		1 decigram (dg) = .1 g		1 deciliter (dL) = .1 L
	1 centimeter (cm) = .01 m		1 centigram (cg) = .01 g		1 centiliter (cL) = .01 L
	1 millimeter (mm) = .001 m		1 milligram (mg) = .001 g		1 milliliter (mL) = .001 L
	1 meter = 100 centimeters		1 gram = 100 centigrams		1 liter = 100 centiliters
	1 meter = 1000 millimeters		1 gram = 1000 milligrams		1 liter = 1000 milliliters

Convert 5 meters into kilometers. From the table we know that 1 kilometer = 1000 meters.

$$5 \; meters \cdot \frac{1 \; kilometer}{1000 \; meters} = \frac{5 \; kilometers}{1000} = .005 \; kilometers$$

Make the following conversions.

k) Convert 12 meters into millimeters.

l) Convert 115 centimeters into meters.

m) Convert 3.2 kilometers to millimeters. (You will first need to convert to meters and then from meters to millimeters.)

Next, lets take a look at a common issue which can occur when working with a mix of units. First, suppose we were adding weights.

$$6 \; lbs \; 12 \; ounces + 2 \; lbs \; 8 \; ounces$$

Adding these, we get:

$$6 \; lbs \; 12 \; ounces + 2 \; lbs \; 8 \; ounces = 8 \; lbs \; 20 \; ounces$$

But there are 16 ounces in a pound. So, we have too many ounces. However, the 20 ounces would be 1 pound and 4 ounces. (It is similar to the idea of having a carry.)

$$8 \; lbs \; 20 \; ounces = 8 \; lbs + 1 \; lb \; 4 \; ounces = 9 \; lbs \; 4 \; ounces$$

Try some on your own.

n) $9\ lbs\ 7\ ounces + 4\ lbs\ 8\ ounces$

o) $8\ feet\ 5\ inches + 9\ feet\ 10\ inches$

p) $5\ minutes\ 45\ seconds + 9\ minutes\ 25\ seconds$

Finally, we could encounter the same issue with multiplication.

$$(3\ feet\ 4\ inches) \times 4$$

If we multiply, we get: $12\ feet\ 16\ inches$. But that is too many inches. Twelve inches make another foot. So, we convert the inches:

$$12\ feet\ 16\ inches = 12\ feet + (1\ foot\ 4\ inches) = 13\ feet\ 4\ inches$$

Try some.

q) $(2\ feet\ 3\ inches) \times 5$

r) $(2\ lbs\ 5\ ounces) \times 4$

s) $(1\ hour\ 35\ minutes) \times 3$

110

Algebra I

Active Lesson: 1.10b

In a final activity reviewing units, we want to work with problems which have mixed units. For example:

Ben needs to cut 55 centimeters from a 2-meter board.

The problem is subtraction but our units don't match. We need to convert one of the units so that they match.

$$55 \; centimeters \cdot \frac{1 \; meter}{100 \; centimeters} = \frac{55 \; meters}{100} = .55 \; meters$$

Now, both units are in meters and we can subtract:

$$2 \; meters - .55 \; meters = 1.45 \; meters$$

Try the following. (Remember, convert one of the units so that they match.)

a) Tim needs to cut 70 centimeters from a 4 meter stick. How long will the stick be after the cut?

b) Sarah can jump 1.78 meters. Her friend can jump 98 centimeters. How much further can Sarah jump than her friend?

We can encounter a similar type of problem where the final answer is requested in a different unit. All we need to do is convert our answer.

There are four boards each 80 centimeters. What is the total length of the boards in meters?

$$4(80 \; centimeters) = 320 \; centimeters$$

The answer was asked for in meters, so we make the conversion.

$$320 \; centimeters \cdot \frac{1 \; meter}{100 \; centimeters} = \frac{320 \; meters}{100} = 3.2 \; meters$$

Try these problems:

c) A carpenter has eight 65-centimeter boards on his truck. He needs to know the total number of meters which he has.

d) A batch of brownies requires 120 ml of milk. Five batches are needed. How many total liters of milk are required?

e) A construction manager has twelve 150-gram concrete blocks. He needs the total number in kilograms. How many kilograms does he have?

In our last lesson, we learned conversions within the U.S. measurement system and within the metric system. Next, we want to convert between the two systems. The process is the same. The conversion numbers are just trickier.

Length		Mass		Capacity	
	1 in = 2.54 cm		1 lb = .45 kg		1 qt = .95 L
	1 ft = .305 m		1 oz. = 28 g		1 fl ounce = 30 mL
	1 yd = .914 m		1 kg = 2.2 lb		1 L = 1.06 quarts
	1 mile = 1.61 km				
	1 meter = 3.28 ft				

Convert 12 feet to meters.

$$12 \ feet \cdot \frac{.305 \ meters}{1 \ foot} = 3.66 \ meters$$

Make the following conversions.

f) Convert 32 feet to meters.

g) Convert 27 pounds to kilograms.

h) Convert .7 kilograms to pounds

(This next problem will require multiple steps. First take centimeters to meters and then meters to feet.)

i) Convert 325 centimeters to feet.

Finally, we will use a formula to convert between two systems. The formulas below can convert from Fahrenheit temperatures to Celsius or from Celsius to Fahrenheit.

$$C = \frac{5}{9}(F - 32)$$

$$F = \frac{9}{5}C + 32$$

Let's convert 72°F into Celsius. We sub in the Fahrenheit temperature and then follow the order of operations.

$$C = \frac{5}{9}(72 - 32)$$

$$C = \frac{5}{9}(40) = 22.\overline{2}°C$$

Work the following:

j)   Convert 55°F to Celsius.

k)   Convert 15°C to Fahrenheit.

We want to begin learning how to solve equations. I will focus on the idea of factors. The book does this in a different order than I would. But let's review the idea of factors. Factors are two numbers that multiply to create another number. Here is an example:

$$3 \cdot 5 = 15$$

In the next section, I will introduce an idea which I call factor puzzles. If we had two numbers multiplied together and we didn't know one, we would have a factor puzzle.

$$3 \cdot x = 15$$

I teach it this way, because it is fairly easy to see that x must be 5. For this activity, we are going to see equations like this:

$$x + 3 = 7$$

The x and 3 are not factors of 7. The addition prevents that. Factors must be two numbers which multiply together to make the number. Circle any of the following which involve factors:

a)

$$3 \cdot y = 21$$

$$z - 2 = 18$$

$$5 + k = 3$$

$$9x = 81$$

Again, we'll take the factor puzzle idea further soon. For now, we want to discover what the missing number is.

$$x + 3 = 7$$

You can probably do this in your head. But, if you couldn't, you simply take away the 3. However, the equal sign means this is like a balance:

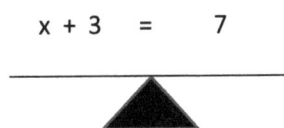

x + 3   =   7

If you remove 3 from one side, it won't remain balanced unless you remove (subtract) 3 from the other side.

$$x \quad = \quad 4$$

Solve these:

b) $y + 7 = 14$

c) $r + 15 = 30$

d) $c + 6 = 18$

And, in the same manner, it can go the other way.

$$x - 5 = 10$$

To discover x, we add 5. But if we do it to one side, we must do it to the other to keep it balanced.

$$x - 5 + 5 = 10 + 5$$

$$x = 15$$

Solve these:

e) $y - 2 = 8$

f) $s - 12 = 20$

g) $a - 7 = 21$

This next set involves negative numbers, but the concept is the same.

$$x + 5 = -2$$
$$x = -2 - 5$$
$$x = -7$$

Work these:

h) $y + 8 = -10$

i) $t - 7 = -12$

j) $b + 3 = -21$

The idea carries over to fractions and decimals too.

$$x + \frac{1}{2} = \frac{3}{4}$$

$$x = \frac{3}{4} - \frac{1}{2}$$

$$x = \frac{3}{4} - \frac{2}{4} = \frac{1}{4}$$

Solve the following:

k) $y + \dfrac{2}{3} = \dfrac{5}{6}$

l) $r - \dfrac{1}{4} = \dfrac{3}{8}$

m) $c + 2.1 = 9.3$

n) $d - 3.4 = 7.1$

o) $x + 4.5 = -2.6$

As we will also discuss more as we work with harder equations, the next step is to "clean-up." This means to combine like terms.

$$5x - 4x = 9$$

$$x = 9$$

They can get more complicated:

$$9x + 5 - 8x - 2 = 12$$

Clean up. (Combine like terms.)

$$x + 3 = 12$$

$$x = 9$$

Try these:

p) $7y - 6y = 14$

q) $8z + 5 - 7z = 9$

r) $9x - 5 - 8x = -12$

s) $7s + 12 - 7 + s - 7s = 2$

t) $2x - 14 + 7 - x = 12 - 6$

This last set of problems requires the distributive property. Clean-up by multiplying and then combining like terms.

$$3(x - 3) - 2x = 7$$
$$3x - 9 - 2x = 7$$
$$x - 9 = 7$$
$$x = 16$$

Work these:

u) $8(x + 1) - 7x = 15$

v) $3(y + 5) - 2y + 7 = -12$

w) $5(v - 2) + 17 - 4v = 12 + 3$

On this problem, remember that the 2 owns the negative sign in front of it. It acts like a -2 as it gets distributed.

x) $3(x + 7) - 2(x - 5) = 3$

With this final problem, clean up both rooms. The variable will disappear from the right side.

y) $4(y - 2) - 3(y + 1) = 5(y + 1) - 5(y + 2)$

Algebra I

Active Lesson: 2.1b

In this activity, we want to begin thinking of math as a language. And, sometimes, we are required to translate between languages. Here, we want to translate between the English language into Math language. Then, we want to use our new algebra skills to solve the equations.

- When you translate, move left to right, just like we do when we read.
- Translate one word at a time.
- Be careful, there are a few translations that can switch up the order. For instance, "2 less than x" would translate as $x - 2$.

Translate and solve.

5 more than x is 11.

Moving left to right, going one word at a time, we have:

> 5: this is already in math language.
>
> More than: this is addition.
>
> x: this is already in math language.
>
> Is: =
>
> 11: this is already in math language.

Putting it all together, our translation looks like this.

$$5 + x = 11$$

a) In the space below, solve for x.

Try some on your own. Translate and solve.

b) 21 more than y is 43.

c) 3 subtracted from x is 12.

d)   14 less than y is 30.

e)   9 more than x is 15.

In these next translations, we can combine like terms.

Translate and solve.

The difference of $7x$ and $6x$ is 5.

$$7x - 6x = 5$$

$$x = 5$$

Work these on your own. Translate and solve.

f)   The difference of $12y$ and $11y$ is 9.

g)   The difference of $110x$ and $109x$ is -32.

We can now use these translation skills to solve some basic word problems. When I do word problems, I start with something I call the dictionary. I'll discuss the dictionary more later. For now, just think of it as a reminder of what we are looking for. (Be sure to give your answer in the context of the problem.)

Two young boys, Max and Sam, weight 152 lbs. If Max weighs 80 lbs, how much does Sam weigh?

I start with the dictionary, which reminds me of what we are searching for.

Math	English
x	Sam's Weight

Translating, I get the following equation:

$$x + 80 = 152$$

Find Sam's weight.

$x = 72$ Sam weights 72 lbs.

Try a few on your own. (I have provided the dictionaries.)

h) Angela and Beth are sisters. Combined, their age is 24. If Angela is 10, how old is Beth?

Math	English
x	Beth's Age

i) Two teams are playing basketball, a red team and a blue team. There are 80 people in the stands. If 65 people are fans of the red team, how many fans are there of the blue team?

Math	English
x	Number of Blue Team's Fans

j) An elementary school teacher has 150 books in her classroom. Some of the books have been donated and the rest were purchased by the teacher. If 90 books were donated, how many were purchased by the teacher?

Math	English
x	Number of Books Purchased by the Teacher

Finally, let's look at some of these basic word problems which involve money.

A shirt was on sale. It was purchased for 16 dollars. The purchase price was 8.50 dollars less than the original price. What was the original price?

Math	English
x	The Original Price of the Shirt

Because this involves "less than" our translation will reverse, and we will have the equation:

$$x - 8.50 = 16$$

The original price was: $x = 24.50$ The shirt originally cost $24.50.

Work these two problems.

k) The local bowling alley charges $5.50 per game on Saturday afternoons. This is $2.75 less than they charge on Saturday nights. How much do they charge for a game of bowling on Saturday nights?

l) An employee of a car dealership gets a discount. They paid $21,450 for their new car. This is $1,575 less than the original price of the car. How much did the car originally cost?

Algebra I

Active Lesson: 2.2a

When I teach algebra, I teach this section before 2.1 because of what I call a factor puzzle. I believe the concept of factors is one of the key ideas in algebra. And factors can help us better understand the process for solving equations.

Here is what I call a factor puzzle:

$$2 \cdot x = 14$$

a) Two numbers are multiplying together to get the number 14. What is the missing number?

b) In algebra, we don't always show the multiplication sign when there is a variable. So, this next problem is multiplication again. Solve the factor puzzle. What value must y equal?

$$5y = 20$$

$$y =$$

Work the following. Find the missing value.

$$8x = 72$$

c) $x =$

$$6z = 36$$

d) $z =$

Try some that involve a negative.

$$-2y = 18$$

e) $y =$

$$-7n = 42$$

f) $n =$

$$-3a = -18$$

g) $a =$

If you aren't sure what the missing number is, all you need to do is divide.

$$11x = 132$$

$$x = \frac{132}{11} = 12$$

Try some that you may not know.

$$24v = 312$$

h) $v =$

$$17x = 306$$

i) $x =$

And if you had one you didn't know which involved a negative sign, just divide by the negative number.

$$-16y = -224$$

j) $y =$

$$-19m = 76$$

k) $m =$

Suppose your puzzle involved division. Some number divided by 7 equals 2.

$$\frac{x}{7} = 2$$

$$x = 14$$

Try these:

$$\frac{y}{6} = 3$$

l) $y =$

$$\frac{z}{5} = 6$$

m) $z =$

But if you ever are unsure of the missing number. Multiply.

$$\frac{y}{12} = 3$$

$$y = 3 \cdot 12 = 36$$

Work these.

$$\frac{x}{15} = 5$$

n) $x =$

$$\frac{s}{16} = 13$$

o) $s =$

Try some that involve negative numbers. Again, if you can't see them in your head, just multiply.

$$\frac{y}{-4} = 8$$

p) $y =$

$$\frac{x}{-22} = -11$$

q) $x =$

Look at this factor puzzle.

$$-m = 15$$

This is really the same as

$$-1m = 15$$

In this case, the value of m must be:

$$m = -15$$

Try some. Remember, if you aren't sure, just divide by -1.

$$-x = 21$$

r) $x =$

$$-r = -7$$

s) $r =$

The next group of problems aren't technically factor puzzles. They involve fractions, and fractions don't divide evenly, which factors must do. However, the concept for solving the missing number remains the same. Just divide.

$$\frac{2}{3}x = \frac{1}{4}$$

So, we are dividing fractions:

$$x = \frac{\frac{1}{4}}{\frac{2}{3}}$$

And when we divide fractions, we Keep, Change, Flip.

$$x = \frac{1}{4} \cdot \frac{3}{2} = \frac{3}{8}$$

Work these. Remember, if your answer is a fraction that can be reduced, do so:

$$\frac{3}{5}y = \frac{2}{3}$$

t) $y =$

u) $\dfrac{4}{3}n = \dfrac{2}{9}$

$n =$

v) $-\dfrac{1}{2}x = \dfrac{1}{2}$

$x =$

w) $-\dfrac{7}{2}s = -\dfrac{14}{5}$

$s =$

Another key idea in solving equations is to "clean up." We will explore this idea more later, but for now, it means to combine like terms. If you need a refresher on combining like terms, return to that activity. Look at this problem.

$$5x + 2x = 21$$

There are terms on the left which are alike and can be combine.

$$7x = 21$$

After we cleaned up the like terms, we have a factor puzzle.

$$x = 3$$

Try some:

x) $5y - 2y = 18$

y) $12m + 2m - 4m = 100$

This problem requires combining like terms on both sides.

z) $-3x + 5x = 8 + 6$

aa) $3y - 8y + 10y = 20 - 25$

On these next problems, the variable is on the right instead of the left. That doesn't change anything. The puzzle remains the same.

bb) $-121 = 14s - 3s$

cc) $15 - 4 + 9 = 18y - 16y + 2y$

Previously, we began translating basic word problems from the English language to Math language. Those previous problems involved addition and subtraction. Here, we want to look at problems which involve multiplication and division. We begin with multiplication.

Translate and solve.

The number 135 is the product of 5 and x.

Here is the translation:

- "135 is" will be translated as $= 135$.
- "the product of 5 and x" will translate as 5x.

Our equation becomes:

$$5x = 135$$
$$x = 27$$

Try this problem.

a) The number 156 is the product of 13 and y.

We can also do division.

x divided by 6 is equal to 9.

$$x \div 6 = 9$$

Or:

$$\frac{x}{6} = 9$$

$$x = 54$$

Work these problems.

b) y divided by 11 is equal to 20.

c) n divided by 4 is equal to -8.

"Quotient of" is still another way to say division. For instance, the quotient of x and 3 would be $\frac{x}{3}$.

Translate and solve these problems.

d) The quotient of x and 7 is 15.

e) The quotient of y and -4 is 14.

We can also work with fractions. For instance, $\frac{1}{3}$ of x means multiply: $\frac{1}{3}x$. So:

$\frac{1}{3}$ of x is 7.

$$\frac{1}{3}x = 7$$

$$3 \cdot \frac{1}{3}x = 3 \cdot 7$$

$$x = 21$$

Try this problem.

f) $\frac{1}{4}$ of x is 23.

Let me show you this next problem.

$\frac{2}{3}$ of y is 18.

$$\frac{2}{3}y = 18$$

Here, it is easiest if we multiply both sides by the fraction turned upside down. (This is called the reciprocal.)

$$\frac{3}{2} \cdot \frac{2}{3}y = 18 \cdot \frac{3}{2}$$

$$y = \frac{3}{2} \cdot 18$$

$$y = 27$$

Work this. Translate and solve.

g) $\frac{3}{4}$ of x is 11.

And, we can have problems involving the addition or subtraction of fractions.

The sum of $\frac{2}{3}$ and x is $\frac{3}{2}$.

$$\frac{2}{3} + x = \frac{3}{2}$$

$$\frac{2}{3} - \frac{2}{3} + x = \frac{3}{2} - \frac{2}{3}$$

$$x = \frac{3}{2} - \frac{2}{3} = \frac{9}{6} - \frac{4}{6} = \frac{5}{6}$$

Work these problems.

h)   The sum of $\frac{1}{3}$ and x is $\frac{13}{12}$.

i)   The difference of y and $\frac{3}{4}$ is $\frac{31}{20}$.

And, as we did before, we can solve basic word problems.

At a candy story, chocolate costs $4 per pound. If someone's bill was $48, how many pounds did they purchase.

The total amount spent would be $4 times the number of pounds purchased: $4x$. And, so our equation becomes:

$$4x = 48$$

$$x = 12$$

Now, we put that into English and the purchased 12 lbs of chocolate.

Try these.

j)   It costs $3.50 for a pound of trail mix. If the total bill was $24.50, how many lbs of trail mix did they purchase?

k)   Three friends plan on splitting their lunch bill evenly. The bill was for a total of $47.25. How much will each friend need to contribute?

Here is a word problem involving "fraction of." I will help you set it up.

A bicycle was on sale for $\frac{4}{5}$ of the original price. If the sale price was $142, what was the original price.

$$\frac{4}{5}x = 142$$

l)   Find the original price.

Algebra I

Active Lesson: 2.3a

Let's return to solving equations. First, we need to get this equation down to a factor puzzle, but the 5 is preventing it.

$$2x + 5 = 15$$

Subtract the 5 from both sides.

$$2x = 15 - 5$$
$$2x = 10$$

Now we can solve the factor puzzle.

$$x = 5$$

Solve. (First, get the problem down to a factor puzzle.)

a) $3y - 7 = 14$

b) $5z + 3 = 28$

c) $6r - 2 = 34$

d) $8x - 5 = -69$

e) $-2y + 4 = -16$

Look at our next problem.

$$5x = 3x + 14$$

We don't have a factor puzzle yet. To make one, we need to get all the x terms back together. In order to do that, we must remove $3x$ from the right. But if we do something to the right, we must do the same thing to the left.

$$5x - 3x = 3x - 3x + 14$$

$$2x = 14$$

Now we have a factor puzzle which we can solve.

$$x = 7$$

Work these:

f) $7y = 3y + 24$

g) $6r = 2r - 12$

h) $8x = x + 49$

In these next problems, to get the variables back together, we need to add instead of subtract.

$$4y = -2y + 18$$

$$4y + 2y = -2y + 2y + 18$$

$$6y = 18$$

$$y = 3$$

136

Solve the following:

i) $7r = -3r + 40$

j) $4z = -2z - 36$

k) $-3x = -2x - 7$

Finally, it isn't necessary to always move the variable to the room on the right. We may need to go the other direction. The process works the same.

$$5y + 8 = 3y$$

If we moved the 3y to the right, it would leave that room empty. Instead, we move the 5y.

$$5y - 5y + 8 = 3y - 5y$$

$$8 = 2y$$

Our factor puzzle is reversed but the concept remains.

$$y = 4$$

Work these:

l) $2x + 4 = 6x$

m) $3c - 12 = 5c$

n) $-7w - 20 = 3w$

o) $-9x + 17 = -10x$

Algebra I

Active Lesson: 2.3b

The next step in solving equations involves moving both variables and constants to the other side.

$$7x + 8 = 5x + 12$$

First, we need to get the x's back together.

$$7x - 5x + 8 = 5x - 5x + 12$$

$$2x + 8 = 12$$

Now we need to get down to a factor puzzle.

$$2x + 8 - 8 = 12 - 8$$

$$2x = 4$$

And solving the factor puzzle is easy.

$$x = 2$$

Work these:

a) $5y + 2 = 3y + 16$

b) $7z - 6 = 2z + 19$

c) $11v - 7 = 2v - 25$

d) $-2x + 1 = -8x + 43$

I believe the idea of a factor puzzle helps us to understand the order in which we are to solve equations. However, as I mentioned previously, the factor puzzle doesn't always involve true factors.

$$3y = 7$$

$$y = \frac{7}{3}$$

Factors are two number that divide evenly, yet we got a fraction for y. Despite this limitation, use the factor puzzle concept to help you work these problems. You will get fractions.

e) $5x + 13 = -2x - 7$

f) $-12r + 4 = -8r - 15$

And, the ideas can extend to fractions and decimals. Use the same process to solve these.

g) $\frac{1}{2}x + 11 = \frac{1}{4}x + 19$

h) $\frac{2}{3}y - 2 = -\frac{1}{6}y + 4$

i) $2.1x + 3 = 1.5x + 9$

j) $1.7y - 5 = 1.8y + 11$

Algebra I

Active Lesson: 2.4a

As the equations become more complicated, the process to solve them is always the same. Follow these three steps for any equation:

1) Clean Up Both Rooms- Do any distributing and combining like terms on both sides of the equation.
2) Get the Variable Back Together- If there are variables on both sides of the equation, move them all to one side.
3) Get the Variable Alone- Get down to a factor puzzle and solve.

Here are a couple of problems which show the steps.

$$3(y - 2) + 7 = y - 9$$

1) Clean Up Both Rooms.

$$3y - 6 + 7 = y - 9$$

$$3y + 1 = y - 9$$

The rooms are now clean. Next, there are variables in both rooms.

2) Get the Variable Back Together.

$$3y - y + 1 = -9$$

$$2y + 1 = -9$$

3) Get the Variable Alone. Make a factor puzzle and then solve.

$$2y = -10$$

$$y = -5$$

Here is another.

$$8(x + 3) + 4 = 2(x - 2) + 2$$

1) Clean Up Both Rooms.

$$8x + 24 + 4 = 2x - 4 + 2$$

$$8x + 28 = 2x - 2$$

2) Get the Variable Back Together.

$$8x - 2x + 28 = -2$$

$$6x + 28 = -2$$

3) Get the Variable Alone. Make a factor puzzle and then solve.

$$6x = -2 - 28$$

$$6x = -30$$

$$x = -5$$

Now, solve some on your own. If you ever don't need a step, just skip it. In this first problem, after cleaning up, there aren't variables in both rooms. Just move on to Get the Variable Alone.

a) $4(r + 5) + 4 = 36$

b) $2(2x - 3) + 7 = 2(x + 5) - 1$

c) $10 - 4(x - 3) = 14x - 14$

d) $2(2y - 1) + 2 = 5(y - 7) + 5$

In this next problem, the solution will be a fraction. However, the process remains the same.

e) $2r + 3(r - 1) + 7 = 2(r + 5) - 8$

This final problem will involve parenthesis inside of parenthesis. Clean up from the inside of the room out.

f) $2[3 - 2(x - 4)] = 6x + 2$

142

Answer if the following are making a mathematical statement which is true or false:

a) $8 = 0$

b) $0 = 0$

c) $2x + 1 = 2x + 1$

d) $3x = 3x - 2$

Equations can fall into three categories: conditional, identity, and contradiction. A conditional equation is one in which there are one or more answers for the variable. This is the typical equation we have solved. They end with a solution to the variable. The next type of equation is called an identity. Follow the steps to solve this equation. It will be an identity. (A strange thing will happen with the variable and that is okay.)

e)
$$4(x - 2) + 13 = 3(x + 1) + x + 2$$

An identity occurs when the variable drops out and what remains is telling the truth. For instance, an equation which ends with $0 = 0$ is an identity. It means that any number makes this equation true, and therefore, the solution is **all real numbers**.

The final type of equation is called a contradiction. Follow the steps to solve this equation. It will be a contradiction. (That strange thing will occur again.)

f)
$$5(2x + 3) - 7 = 2(5x + 1) - 2$$

A contradiction occurs when the variable drops out and what remains is *not* telling the truth. For instance, an equation saying $8 = 0$ is a contradiction. It means that there are no numbers which can make this equation true, and therefore, there is **no solution**.

Solve the following equations and give the solution. If there are infinite solutions or no solutions, indicate as such. Also, write whether or not the equation should be classified as conditional, an identity, or as a contradiction.

g) $5x - (x + 2) = 3x + 2(x + 4) - x$

h) $6(3x + 5) = -2(x + 15)$

i) $12 + 6(2x - 5) = 4x + 2 + 2(4x - 6) - 8$

Algebra I

Active Lesson: 2.5a

Add the following fractions:

a) $\dfrac{1}{2} + \dfrac{1}{3} =$

b) $\dfrac{3}{4} + \dfrac{1}{6} =$

c) To add fractions, what must you do with the denominator?

The fancy name for this is called the LCD, least common denominator. And the LCD has a useful trick in solving equations with fractions.

$$\frac{1}{2}x + \frac{1}{3} = \frac{5}{6}$$

If we look for an LCD for the fractions involved in this equation, we can see that it is 6. Here's the trick. Multiply every term in this equation by 6. (A term is separated by a + or − or =.)

$$6 \cdot \frac{1}{2}x + 6 \cdot \frac{1}{3} = 6 \cdot \frac{5}{6}$$

The reason this is helpful is that all of the fractions will be removed from the equation.

$$3x + 2 = 5$$

d) What is the solution?

This trick is legal because we have multiplied both sides of the equation by the same number, so the equation hasn't been changed.

Try some on your own.

e) $\dfrac{2}{3}y - \dfrac{1}{2} = \dfrac{1}{6}$

f) $\dfrac{1}{8}x + \dfrac{1}{4} = \dfrac{1}{2}$

The idea doesn't change if we have more terms. Multiply the LCD by each term.

g) $\dfrac{2}{3}x - \dfrac{1}{4} = \dfrac{1}{2}x + \dfrac{1}{12}$

On this next problem, you will get a fraction for your answer.

h) $-\dfrac{1}{5}y + \dfrac{1}{10} = \dfrac{3}{5}y + \dfrac{3}{10}$

Look at this problem:

$$4 = \frac{1}{3}(x - 2)$$

Because of the parenthesis, there is only one continuous group on the right side. We will multiply both sides by 3 and get this:

$$3 \cdot 4 = 3 \cdot \frac{1}{3}(x - 2)$$

$$12 = x - 2$$

$$x = 14$$

Try a couple.

i) $8 = \frac{1}{2}(y + 1)$

j) $3 = \frac{2}{3}(a - 3)$

Try extending the idea here. (Multiply both sides by 12.)

k) $\frac{1}{4}(x + 2) = \frac{1}{3}(x + 3)$

These next problems will look a bit different, but they aren't. There is one group on each side. The LCD is 6.

$$\frac{2y + 4}{3} = \frac{y - 2}{2}$$

$$6 \cdot \left(\frac{2y + 4}{3}\right) = 6 \cdot \left(\frac{y - 2}{2}\right)$$

$$2(2y + 4) = 3(y - 2)$$

l)   Finish the problem from here.

Try another.

m)

$$\frac{3x - 2}{4} = \frac{3x + 5}{8}$$

Algebra I

Active Lesson: 2.5b

We can follow a similar approach to our last activity when working with equations with decimals.

$$.02x + .03 = .05x - .03$$

Look through the problem at the decimals. Find the furthest decimal place. In this problem they all go to the second decimal place—the hundredths. Now, multiply every group in the equation by 100. Remember, groups are separated by + or − or =.

$$100(.02x) + 100(.03) = 100(.05x) - 100(.03)$$

After we multiply through, the decimals will be removed.

$$2x + 3 = 5x - 3$$

a) Now we have a much easier equation to solve. In the space below, finish solving it.

Try one on your own. Notice that again the furthest decimal place is the hundredths so we will multiply by 100.

b) $.1 + .02y = .03y + .05$

In this next problem, there are only three groups. The $-.02(y + 2)$ is all one group because of the parenthesis. I've started this problem for you.

$$.25y - .02(y + 2) = .42$$

$$100 \cdot (.25y) - 100 \cdot (.02(y + 2)) = 100 \cdot (.42)$$

$$25y - 2(y + 2) = 42$$

c) Finish this problem.

Work this problem.

d) $.25x + .05(x - 3) = .75$

In these final two problems, we have decimals with different lengths. Remember, multiply through by the furthest decimal.

e) $.02x - .5 = .04x + .12$

f) $.01y + .15 = .2(y - 4)$

Algebra I

Active Lesson: 2.6a

a) Circle any of the following which are rates (speeds).

$$55 \ miles \ per \ hour$$

$$500 \ miles$$

$$15 \ miles \ per \ hour$$

$$12 \ hours$$

b) Suppose you are driving your car at 60 mph. If you have driven at that speed for two hours, how far have you gone?

Because you are familiar with a situation like this, the math isn't too hard. But what we are really doing in this problem is using an important formula.

$$distance = rate \cdot time$$

Which is typically abbreviated as:

$$D = R \cdot T$$

In the problem I started with, we had the rate 60 mph, and the time 2 hours.

$$D = 60 \cdot 2$$

So, we could find the distance. Try another. Find the distance.

$R = 45 \ mph$

$T = 3 \ hours$

c) $D =$

This is straightforward. However, suppose we had the distance and time and needed the rate. We could use our algebra skills.

Suppose you drove 450 miles in 9 hours, what was your average speed. The set up would look like this:

$$450 = R \cdot 9$$

This isn't the normal order we use for algebra problems, but the order we multiply doesn't matter, so it can be written as:

$$450 = 9R$$

d) Solve this:

Try a couple on your own.

e) You drove 310 miles in 5 hours, what was your average speed?

f) You drove 437.5 miles in $6\frac{1}{4}$, what was your average speed? (You could work with fractions here, but I would just change $6\frac{1}{4}$ into 6.25. It is fine to use a calculator on these.)

In some problems, you will have the distance and the rate. The concept is the same, just use your algebra skills to find the time. Solve a couple of these.

g) You drove 330 miles at 55 miles per hour. Find the time you travelled.

h) You drove 338 miles at 42.25 miles per hour. Find the time you travelled.

Algebra I

Active Lesson: 2.6b

Algebra is often used in courses like Physics and Chemistry to solve formulas. However, it is often done with only variables. Here's what I mean.

$$d = r \cdot t$$

Solve this formula for r.

$$\frac{d}{t} = r$$

No numbers; just variables. The trick is to follow the same process of algebra. Pretend every other variable were a number.

$$9 = r \cdot 5$$

I would move the r to the back so that it looks like what we are used to.

$$9 = 5r$$

Here, to get r alone, we would divide by 5.

$$\frac{9}{5} = r$$

I only do this to follow the pattern that I would need. Try this:

a) Solve for t.

$$d = r \cdot t$$

Next, let's work with the formula $A = \frac{1}{2}bh$. This is the area of a triangle.

We will solve the formula for b. I'll help you start. I'd move the b to the back so it is like what we are used to.

$$A = \left(\frac{1}{2}h\right)b$$

Now, treat the circle like a number in front of a variable. We would divide it.

$$\frac{A}{\frac{1}{2}h} = b$$

As you are first learning how to do this, I would accept this as an answer. However, a fraction on the bottom turns upside down and we would get:

$$\frac{2A}{h} = b$$

b)   Now you try. Solve $A = \frac{1}{2}bh$ for h.

This next formula is called the simple interest formula $I = Prt$.

c)   Solve the formula for P. (Hint, put the P at the back and then move the rt.)

d)   Try another. Solve the formula $P = a + b + c$ for a. (Imagine everything else is a number. You would subtract to get a alone.)

e)   Next, solve the formula $P = a + b + c$ for b.

For us, the most significant application of this idea involves working with lines. We'll discuss the idea in depth later, but for now we want to use this idea to solve an equation for y.

$$\boxed{2x} + y = 5$$

f)   Act like the circle is all one number. We would subtract it to get y alone. Finish below.

g) Try another. Solve for y.

$$4x + y = 9$$

Let's add a coefficient in front of the y.

$$\textcircled{3x} + \underline{2y} = 5$$

We need to get rid of the $3x$ and the 2. If we follow our normal process, we need to get down to the factor puzzle, so the 3x would leave first.

$$2y = 5 - 3x$$

Then to get rid of the 2, we must divide it on the other side. However, we must divide the entire side.

$$y = \frac{5 - 3x}{2}$$

h) Work these on your own. Solve for y.

$$2x + 5y = 7$$

i) Solve for y.

$$7x + 9y = 5$$

Algebra I

Active Lesson: 2.7a

To begin, I want us to look at the key ideas behind inequalities. Look at the following:

$$5 + 3 < 12$$

a) Is this true?

In the space below, subtract 3 from each side.

b) Is the inequality true now?

Next, try it again with a slight difference.

$$5 - 3 < 12$$

c) In the space below, add 3 to both sides.

d) Is the inequality true now?

Now, look at this inequality.

$$2 \cdot 6 > 8$$

e) Is the inequality true?

f) In the space below, divide both sides by 2.

g) Is the inequality true now?

Try this one.

$$\frac{6}{2} > 2$$

h) Is this inequality true?

i) In the space below, multiply both sides of the inequality by 2.

j) Is this new inequality true?

If we add or subtract on both sides of an inequality, it remains true. If we multiply or divide on both sides of an inequality by a positive number, it remains true. Now, let's see why something goes wrong if we multiply or divide an inequality by a negative number.

$$16 > 10$$

k) In the space below, divide both sides by -1, but leave the inequality sign the same as it is now.

l) Is this new inequality true?

Let's see why this is. Imagine our inequality on a scale.

$$16 \; > \; 10$$

The number 16 is heavier than the number 10. Now, I've changed both sides to their opposite.

$$-16 > -10$$

But this isn't true anymore. Taking the opposite of both sides will not be correct unless we also take the opposite of the inequality.

$$-16 \; < \; -10$$

So, if we multiply (or divide) a negative across an inequality, we must flip the inequality sign. But otherwise, the steps in solving inequalities are the same as solving equations. Solve the following:

m) $2x - 3 > 5$

(On this problem, you will subtract, but that doesn't require flipping the inequality.)

n) $5y + 3 < 18$

(Here you will divide by a negative, so you'll need to flip the inequality.)

o) $-3c - 6 \leq 21$

p) $-5x + 9 \geq -46$

Anytime an equation involves an inequality, the answer isn't a single number, it is a range of numbers. The most common way to show this is with something called interval notation. Suppose we found an answer of:

$$x > 2$$

On a number line, this would be:

The parenthesis means that we don't want to include the 2 because we had a greater than. Interval notation goes up the number line from left to right. (Just like we read a book.) Our interval starts at the number 2 and goes forever. Forever is infinity. Here is the interval notation.

$$(2, \infty)$$

Put the following in interval notation.

q) $x > 9$

r) $x > -3$

Next look at this inequality.

$$x \geq 2$$

Here we want to include the 2. We use a [. This is the number line.

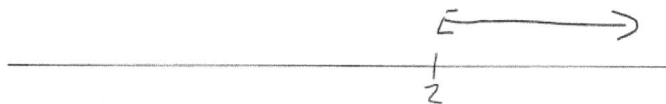

The interval notation would be:

$$[2, \infty)$$

Put the following in interval notation.

s) $x \geq 9$

t) $x \geq -3$

We could also go the other direction.

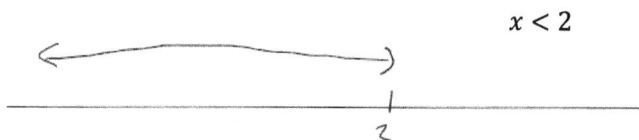

$$x < 2$$

The numbers we want starts at the negative end of the number line. This is negative forever, or negative infinity.

$$(-\infty, 2)$$

Here is the interval notation for $x \leq 2$.

$$(-\infty, 2]$$

We can never reach infinity or negative infinity, so it will never have a bracket. It will always have a parenthesis. Put the following in interval notation.

u) $x < 9$

v) $x \leq -3$

Anytime we have a solution with an inequality, we will add interval notation. Go back to the problems you solved at the beginning of this activity and add the appropriate interval notation.

Solving equations with inequalities can get more complicated, however it always follows the three steps we learned previously:

1) Clean-up
2) Get the variable together
3) Get the variable alone.

Here are some more difficult inequalities. Solve and put the answers in interval notation.

a) $7x - 3 < 4x + 18$

b) $15 - 2y \geq 7y - 21$

c) $4(x - 3) + 15 > -3x + 24$

d) $8r - 2(r - 6) \leq 4r + 2$

Sometimes the variables will "fall out" of an equation. If that occurs, you may be left with an inequality statement that isn't true. If so, we say "no solution." **No Solution** can't be put in interval notation. Work this example.

e)

$$3x - 15 > 3(x + 7)$$

Likewise, sometimes the variable "falls out" and what is left is true. In this case, we say the solution is "all real numbers" or "infinite solutions." In interval notation, **All Real Numbers** would be $(-\infty, \infty)$. Work this example.

f)

$$6x + 14 \geq 2(x - 3) + 4x$$

We can also have inequality problems where it is best to clear out the fractions. The process is identical to what we did previously. Multiply each group by the LCD. Work this problem. The LCD here is 12. (Don't forget to put your answer in interval notation.)

g)

$$\frac{1}{12}x - \frac{2}{3} \geq \frac{1}{6}x + \frac{1}{4}$$

Algebra I

Active Lesson: 3.1a

a) Translate the following, using the dictionary to the right.

Hola. Como estas. Me gusta perros.

Spanish	English
Como	How
Estas	Are You
Gusta	Like
Hola	Hello
Me	I
Perros	Dogs

Several times, I have referred to mathematics as a language. And, sometimes, we are tasked with translating between English and into math language. We follow the same process we would do above. Here's another.

b) Soy de Los Estados Unidos. De donde eres?

Spanish	English
De	From
Donde	Where
Eres	Are You
Estados	States
Los	The
Soy	I am

One word at a time, we work from left to right, translating from one language to another. And, as we will see, using a dictionary in mathematics will be helpful too.

Let's see the process with a word problem:

Trina bought a bicycle that was on sale for $100, which was four-fifth the regular price. What was the regular price?

First, make a dictionary. Our dictionary doesn't need to be every word, but it needs to tell us what we are looking for.

Math	English
x	Original Price

The variable x will be the original price, which we currently don't know. Next, we will translate into math language. There are a lot of words, but I've underlined what needs to be translated.

Trina bought a bicycle that was on sale <u>for $100, which was four-fifth the regular price.</u> What was the regular price?

Now I'll translate, left to right.

$100: this is already in math language.

Was: this is an = sign

Four-fifth's the regular price: this is four-fifth times the original price, which I don't know. $\frac{4}{5} \cdot x$

Put it all together.

$$100 = \frac{4}{5}x$$

Finish solving for x.

$$\frac{5}{4} \cdot 100 = \frac{5}{4} \cdot \frac{4}{5}x$$

$$125 = x$$

Lastly, use the dictionary to put the answer back into English.

Math	English
x	Original Price

The original price of the bicycle is $125.

Now, you try. I'll help with the set up.

The sale price of a guitar is $220. This is two-thirds of the original price. What was the original price?

First, our dictionary.

Math	English
x	Original Price

Second, translate into math language. I've underlined what needs to be translated.

The sale price of a guitar <u>is $220. This is two-thirds of the original price.</u> What was the original price?

c) Create and solve an equation in the space below.

d) Finally, use your dictionary to tell what your answer means in English.

164

Here's another.

A high school play sold 120 student tickets. The number of student tickets sold was twice the number of adult tickets sold.

Here's the dictionary.

Math	English
x	Adult Tickets

I know the number of student tickets. I don't know the number of adult tickets. Next, let's translate.

A high school play sold <u>120 student tickets. The number of student tickets sold was twice the number of adult tickets sold.</u>

120

Was: =

Twice the number of adult tickets: 2x

$$120 = 2x$$

$$x = 60$$

Using our dictionary to put this back into English, we have: The play sold 60 adult tickets.

Try this one on your own. It is a bit harder.

A high school play sold 253 student tickets. The number of student tickets sold was three more than twice the number of adult tickets sold.

e)   Make your dictionary.

f)   Translate and solve.

g)   Put it back into English.

Work these on your own.

h)   A student has 13 notebooks in their locker. The number of notebooks is three more than twice the number of textbooks. How many textbooks do they have?

i)   A gardener has 25 roses in their garden. The number of roses is 4 more than three times the number of tulips. How many tulips are in the gardener's garden?

Algebra I

Active Lesson: 3.1b

Much of this chapter will be devoted to word problems. In this activity, we will work some new variations. We will continue to use our ideas of translation and the dictionary. The first type of problems we will see are number problems.

The difference between a number and 7 is 24.

First, our dictionary.

Math	English
x	Missing Number

Next, we translate one word at a time.

Difference between: Subtraction

A number: x

7: already in math language

Is: =

24: already in math language

We need to be careful with subtraction when we translate. It can prevent us from simply moving left to right across the sentence. However, "difference between" isn't a problem. The difference between x and 7 looks like this:

$$x - 7 = 24$$
$$x = 31$$

The missing number is 31.

Try one.

a)   The difference between a number and 51 is 12.

Here's the same idea, but the missing number is now the second in the subtraction.

b) The difference between 17 and a number is 12.

Students will often want to do these problems in their head, but even the simple problems should be done by algebra. The goal is to gain the skills to work the more difficult problems which are coming. Here is a slightly harder number problem.

The sum of three times a number and 12 is 45.

The dictionary remains the same.

Math	English
x	Missing Number

Translating one word at a time.

Sum: +

Three times a number: 3x

12: 12

Is: =

45: 45

Here, sum of, takes our translation a bit out of order.

$$3x + 12 = 45$$

c) Finish the problem.

Try one.

d) The sum of twice a number and 17 is 67.

Next, we will look at problems which involve two unknown numbers.

One number is seven more than another number. The sum of the numbers is 77.

The key is that there is only one complete unknown, x, for the second number builds off the first. I've added a second number to my dictionary.

Math	English
x	Missing Number
x+7	2nd Number

Notice that the second number is built off the first number.

We've already done some of the translating by making the two numbers. Now we finish it.

The sum of the numbers is 77.

$$x + (x + 7) = 77$$

I've added the second number in parenthesis to show that it is one entire number. At this point, the parenthesis won't change the answer, but when the problems get harder, you might need to multiply.

To solve, we clean up. Without any multiplication, we can ignore the parenthesis and combine like terms.

$$2x + 7 = 77$$

$$2x = 70$$

$$x = 35$$

Putting the answer back into English is a little different here. There are two numbers. I use the dictionary to help. I know the first number. The second number is seven more.

Math	English
x	Missing Number
x+7	2nd Number

The first number was 35. The second number was 42. $(35 + 7 = 42)$

Work this problem.

e) One number is 15 more than another number. The sum of the numbers is 57.

Try another. Here, the sum of the numbers will be a negative number. This would be hard to do in your head. It will be much easier with algebra; just follow the steps we've been using.

f) One number is 6 more than another number. The sum of the numbers is -32.

Here, follow the same ideas, but one of the numbers will be less than. I'll provide you with the dictionary.

The sum of two numbers is 93. One number is 5 less than the other.

Math	English
x	Missing Number
x-5	2nd Number

g) Finish the problem.

Next, one of the numbers involve a multiple of the other number.

The sum of two numbers is 55. One number is 4 more than twice the other number.

Here's the dictionary.

Math	English
x	Missing Number
2x+4	2nd Number

Here's the equation.

$$x + (2x + 4) = 55$$

I've put the parenthesis to show the entire second number. However, there is no further multiplication
h) so just combine like terms to solve. Finish the problem below.

Try one on your own.

i) The sum of two numbers is 66. One number is two more than three times the other number.

Our next type of problem will be "consecutive" number problems. They work just like the number problems we've been doing. However, building the second number is based on the idea of being "consecutive." Here's an example.

The sum of two consecutive integers is 27. What are the numbers?

To build our dictionary, we need the idea of consecutive. It means next to each other. So, if one number is "I don't know" then the second number is "one more than I don't know."

Here's our dictionary.

Math	English
x	1st Number
x+1	Consecutive Number

Now, we just add these two numbers.

$$x + (x + 1) = 27$$

j) Solve for both numbers.

Try one on your own.

k) The sum of two consecutive integers is 91. What are the numbers?

Try another. Here the numbers sum to be a negative number.

l) The sum of two consecutive integers is -45. What are the numbers?

Now, we extend the idea to three consecutive integers.

The sum of three consecutive integers is 99. What are the numbers?

Here's the dictionary. Notice how the third number is now two more. All three numbers must be in a row.

Math	English
x	1st Number
x+1	2nd Number
x+2	3rd Number

The equation would look like this:

$$x + (x + 1) + (x + 2) = 99$$

Then, we combine like terms:

$$3x + 3 = 99$$

m) Finish solving. Remember, you must state what all three numbers are.

Work this problem.

n) The sum of three consecutive integers is -24. What are the numbers?

Finally, we want to work with consecutive even integers. Look at the number line below.

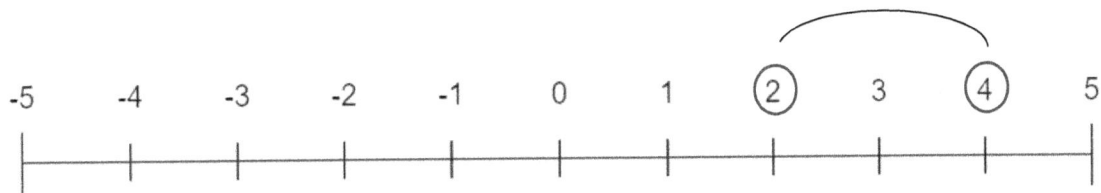

Before, consecutive numbers needed us to add 1. But here, consecutive even would need us to add 2, because there are two steps between any pair of even numbers.

The sum of two consecutive even integers is 56. What are the numbers?

Here's the dictionary.

Math	English
x	1st Number
x+2	2nd Number

And, here's our equation.

$$x + (x + 2) = 56$$

0) Finish solving.

Work this problem.

p) The sum of two consecutive even integers is -84. Find the integers.

Work one last problem. It involves the sum of three consecutive even integers. I'll provide you with the dictionary.

The sum of three consecutive even integers is 96. What are the integers?

Here's the dictionary.

Math	English
x	1st Number
x+2	2nd Number
x+4	3rd Number

q) Explain why the third number is $x + 4$?

r) Build your equation and finish solving. Remember, you are giving three numbers as your answer.

Look at this proportion problem.

$$\frac{2}{3} = \frac{x}{9}$$

You could solve this by getting a common denominator on both sides, but the method you probably were taught for proportions involves lassoing.

Lasso diagonally, multiplying the numbers in each lasso.

$$2 \cdot 9 = 3 \cdot x$$

$$18 = 3x$$

Then the equation is quite simple.

$$x = 6$$

Solve the following proportions using the lasso method.

a) $\quad \dfrac{4}{7} = \dfrac{x}{35}$

b) $\quad \dfrac{9}{15} = \dfrac{3}{x}$

In this activity, we'll be solving problems involving percentages. To do so, we'll be setting up proportions and using the lasso idea. I believe the easiest way to solve proportion problems is to set them up using this pattern:

$$\frac{is}{of} = \frac{\%}{100}$$

Problems of this type will give you two of the three: is, of, %. You then use the lasso technique to find the one which is missing.

For example:

What percentage of 55 is 45?

What percentage: % is missing, so it will be x.

Of: 55

Is: 44

We substitute into our equation:

$$\frac{44}{55} = \frac{x}{100}$$

Then we lasso:

$$44 \cdot 100 = 55 \cdot x$$

And solve:

$$4400 = 55x$$

$$x = 80$$

Remember, x was our missing percentage, so our answer is 80%.

Try one.

c) What percentage of 40 is 10?

Try another. This time you will get a decimal. Round to one decimal place.

d) What percentage of 62 is 35?

These problems simply vary what is missing. Here's an example.

10% of 75 is what number?

%: 10

Is: Missing. We'll make this x.

Of: 75

So, our equation becomes:

$$\frac{x}{75} = \frac{10}{100}$$

e) Finish solving.

Try these. (If your answer has a decimal, round to one decimal place.)

f) 30% of 82 is what number?

g) 20% of what number is 42? (Notice that here you are missing "of.")

In more complex word problems, it may be a little harder, but we still want to find "%", "is", and "of." Let's look at this example.

A couple wants to leave a 15% tip at a restaurant. The bill was $82. What is the tip?

Here's our key pieces.

%: 15

Is: We don't know. This is the tip and it will be x.

Of: 82. (It is harder here, but this is the total bill.)

So, we set up our equation.

$$\frac{x}{82} = \frac{15}{100}$$

h) Finish solving. (This is money, so give your answer with two decimal places.)

Work this problem.

i)   The bill at a restaurant was $105. What should the tip be if they want to leave 20%?

In this problem, we have the % and "is" but not "of."

j)   A protein bar has 12 grams of protein. That is 65% of the daily requirement of protein. Find the daily requirement. (The daily requirement would be our missing "of.")

Finally, this problem will have the "is" and the "of" but not the "%."

k)   An ice cream cone has 350 calories. 185 of those calories are from saturated fat. What percentage of the ice cream is saturated fat?

Algebra I

Active Lesson: 3.2b

Solve the following using the proportion method we did in the last activity:

a) 18 is what percentage of 45? (Round to one decimal place if necessary.)

In this activity, we'll be looking at percent increase or percent decrease. These problems are based on the proportion problems we did in the last section with only a couple of extra steps. Our formula for solving the percentage problems was:

$$\frac{is}{of} = \frac{\%}{100}$$

Here we will modify the formula.

$$\frac{change}{original\ price} = \frac{\%}{100}$$

Change is how the price has changed. We always want it to be a positive, so our formula becomes:

$$\frac{|new\ price - original\ price|}{original\ price} = \frac{\%}{100}$$

Let's work through an example:

A t-shirt is on sale for $12. The original price was $15. Find the percent discount.

Plugging into the formula we have:

$$\frac{|12 - 15|}{15} = \frac{x}{100}$$

We don't know the percentage, so it is x. Notice that there is an absolute value:

$$\frac{|-3|}{15} = \frac{x}{100}$$

After subtracting, the absolute value does its power and we no longer need it.

$$\frac{3}{15} = \frac{x}{100}$$

Finally, we lasso and solve just as we did before.

$$3 \cdot 100 = 15 \cdot x$$

$$300 = 15x$$

$$x = 20$$

Percent increase and percent decrease problems work the same, so in our final answer we need to indicate if this was an increase or decrease. The shirt went on sale; therefore, this was percent decrease.

The percent decrease was 20%. (Or, the shirt was 20% off.)

Try a couple on your own. (Round to one decimal place if necessary.)

b) A jacket has been reduced in price from $55 dollars to $48. Find the percent decrease.

c) The price of gas has dropped from $3.20 per gallon to $3.00 per gallon. What is the percentage decrease?

The concept is exactly the same for percentage increase.

d) Two years ago, the average salary at a local company was $38,000. Now, the average salary is $42,000. Find the percentage increase.

Our next idea comes from finance. When you invest your money, the most basic way in which the bank can pay you is called simple interest. The idea is that at the end of your investment, they simply give you the interest that you earned. (Other methods of paying interest are more complex, and make you more money!)

The formula to calculate your interest using this simple approach is:

$$I = Prt$$

Where:

$I = Interest$- how much money you made.

$P = Principal$- the amount you started with.

$r = rate$- the percentage rate the bank is giving you for your money. (Use the rate as a decimal.)

$t = time$- the time (usually in years) that the bank has your money.

These problems simply require plugging in the numbers into the correct spot in the formula. Let's look at one.

You invested $5000 in the bank for 3 years at 5%. How much simple interest did you earn?

$$I = Prt$$

$I = this\ is\ our\ variable$

$P = \$5000$

$r = .05$ Be sure to change it to a decimal. Move the decimal two places to the left.

$t = 3$

Substitute in the values. (Notice, I put them in parenthesis, so that the multiplication is not lost.)

$$I = (5000)(.05)(3)$$

Now, solve for the interest by multiplying.

$$I = \$750$$

Try this problem on your own.

e) $10000 was invested in a local bank for 6 years. The bank gives an interest rate of 6.5%. Find the simple interest.

In algebra, you won't always be solving for interest, so you will need your algebra skills. In this next problem, we need to find the principle.

An investment made $450 in interest after being in the bank for 5 years at an interest rate of 8%. How much money was originally invested at simple interest?

$I = 450$

$P = this\ is\ our\ variable$

$r = .08$

$t = 5$

Next, we plug into our equation.

$$450 = P(.08)(5)$$

Simplify and then use algebra to solve.

$$450 = .4P$$

$$P = \$1125$$

Try these:

f)   A student made $2250 interest. The money was in a CD for 8 years with an interest rate of 8.5%. How much money was originally invested at simple interest?

In this problem, you will be missing r. (When you get your answer, it will be a decimal. Move the decimal two spaces to the right to turn it into a percentage.)

g) $10,000 was invested into a college savings fund offering simple interest. At the end of 10 years, the fund had earned $1,150. What was the annual rate of return?

Algebra I

Active Lesson: 3.2c

In this activity, we will work with the ideas of percent discount or percent mark-up. I do this in a way that is the reverse of the book because I find it much easier. Here's the first idea that we need.

For percent discount, we need to see the connection with the new price.

If something is 40% off, the new price is 60% of the original price.

If something is 25% off, the new price is 75% of the original price.

Answer the following:

a) A shirt is 10% off, the new price is ___% of the original price.

b) A coat is 30% off, the new price is ___% of the original price.

c) A skirt is 18% off, the new price is ___% of the original price.

These types of problems will ask the sale price of the item and how much money it has been reduced. Here's an example.

A shirt is 15% off the original price of $55.

a) What is the sale price?

The new price of the shirt is $100\% - 15\% = 85\%$ of the original price. So, the new price is:

$$55(.85) = \$46.75$$

b) How much was it reduced? That is just the difference between the new price and the old price.

$$\$55 - \$46.75 = \$8.25$$

Try these. (Since it is money, round to two decimal places if necessary.)

A pair of pants is 24% off the original price of $65.

d)   a)   What is the sale price?

e)   b)   How much were the pants reduced?

The original price for a set of tires was $325. They were marked down 15%.

f)   a)   What is the sale price of the tires?

g)   b)   How much were the tires reduced?

We can extend the same idea to percent mark-up.

If something is marked up 40%, the new price is 140% of the original price.

If something is marked up 25%, the new price is 125% of the original price.

Answer the following:

h)   A shirt is marked up 10%, the new price is ___% of the original price.

i)   A coat is marked up 30%, the new price is ___% of the original price.

j)   A skirt is marked up 18%, the new price is ___% of the original price.

These types of problems will ask the marked-up price of the item and how much money it has changed. Here's an example.

A shirt is marked up 15% from the original price of $55.

    a) What is the marked-up price?

The new price of the shirt is $100\% + 15\% = 115\%$ of the original price. So, the new price is:

$$55(1.15) = \$63.25$$

(Notice that 115% as a decimal moves two spots to the left and becomes 1.15.)

    b) How much was it reduced? That is just the difference between the new price and the old price.

$$\$63.25 - \$55 = \$8.25$$

(Notice that I put the higher price first.)

Try these. (Since it is money, round to two decimal places if necessary.)

A pair of pants is marked up 14% from the original price of $65.

k)   a) What is the new price?

l)   b) How much were the pants marked-up?

The original price for a set of tires was $325. They were marked up 6%.

m)   a) What is the marked-up price of the tires?

n)   b) How much were the tires marked-up?

There is one other type of problem which you may see in this section. It will ask for the discount rate for an item on sale. But a discount rate is exactly the same thing as a percent discount, which we've already done.

$$\frac{|new\ price - original\ price|}{original\ price} = \frac{\%}{100}$$

Work the following problem.

A shirt is on sale for $18.50. The original price was $25.

o)   a) Find the discount rate. (This is the % in the formula above.)

p)   b) How much was the shirt discounted? (This is $|new\ price - original\ price|$)

Algebra I

Active Lesson: 3.3a

Work the following number problem.

a)   One number is five more than twice another number. Their sum is 29.

Math	English
	1st Number
	2nd Number

In this section, we will be working with number problems which adds an extra step. First, answer the following:

b)   If I have 5 quarters, how much money do I have?

c)   If I have 12 dimes, how much money do I have?

d)   If I have 7 nickels, how much money do I have?

The secret, of course, to these problems is that we need to consider the value of each coin. That is the extra twist which we will see in this first group of number problems.

You have a bag filled with quarters and nickels. There are seven more nickels than quarters. How many of each coin do you have if the value of the coins is $1.85?

As always, we build our dictionary.

Math	English
	# Nickels
	# Quarters

Notice, I have the number of each coin in the dictionary. That is what we are looking for. We can build nickels from quarters, and so:

Math	English
x+7	# Nickels
x	# Quarters

Suppose we tried to add the coins to get the total:

$$x + (x + 7) = 1.85$$

But this doesn't work. What is on both sides must be the same kind of thing. On the left is the total number of coins. On the right is the total amount of money. So, we have an extra step. We need to add the value of each coin.

$$.25x + .05(x + 7) = 1.85$$

Now we have the value of the coins on the left and the total value on the right. You could then work with the decimals if you like, but I prefer to work the trick where we eliminate them. I would multiply each group in the problem by 100.

$$25x + 5(x + 7) = 185$$

$$25x + 5x + 35 = 185$$

$$30x + 35 = 185$$

$$30x = 150$$

$$x = 5$$

Looking at our dictionary, x is the number of quarters. However, we need both the number of quarters and the number of nickels. So:

Math	English
12	# Nickels
7	# Dimes

Try one on your own.

e) You have $1.95 in quarters and nickels. You have three more nickels than quarters. How many of each coin do you have.

Math	English
	# Nickels
	# Quarters

f) Set up your equation and solve. Remember, you must multiply the number of the coins by their value so that the left side and the right side are both talking about the value of the coins.

g) Finally, state the number of *each* coin that you have.

Try another. Notice, that this problem is about nickels and dimes, so the values of your coins have changed. Otherwise, the concept is the same.

h) You have $2.35 in nickels and dimes. You have thirteen more dimes than nickels. How many of each coin do you have?

Math	English
	# Nickels
	# Dimes

Work one more coin problem. I'll help you by providing the more difficult dictionary.

You have a pocket full of nickels and dimes. The number of dimes is three more than twice the number of nickels. The value of the coins is $4.05. Find the number of nickels and dimes.

I've built the dictionary. The number of dimes can be built from the number of nickels, so nickels is x.

Math	English
x	# Nickels
2x+3	# Dimes

i)   Set up and equation and solve.

j)   Give the number of each coin. Use the dictionary to help. You now have x. Multiply x by 2 and then add 3 to find the number of dimes.

This idea is exactly the same when working with stamps. Answer the following:

k)   You have twelve 15 cent stamps. What is their value?

l)   You have five 21 cent stamps. What is their value?

m)   As we saw with money, we multiply the number of stamps by how much they are worth. Try a problem.

You have .27 cent stamps and .15 cent stamps. The number of .15 cent stamps is five more than twice as many .27 cent stamps. If the total value of the stamps is $4.74 find the number of each type of stamp.

Math	English
	# .27 stamps
	# .15 stamps

We'll end with this concept extended to the sale of tickets.

A school play sold adult and child tickets to their school play. The number of children's tickets sold was 16 more than twice the number of adult tickets. Children's tickets cost $6 each and adult tickets cost $10 each. If the school sold $5046 worth of tickets, how many of each type of ticket did they sell?

This is identical to what we've been doing. Here's the dictionary:

Math	English
2x+16	# children's tickets
x	# adult tickets

Now, just as we did before, to create our equation, we'll multiply by the value of each ticket.

$$6(2x + 16) + 10x = 5046$$

n) Finish solving this equation and be sure to give the number of *each* type of ticket.

Here is a variation of the ticket problem.

A local play sold 550 tickets. Adult tickets cost $12 each and children's tickets cost $7. If they sold $4750 worth of tickets, how many of each type of ticket did they sell?

What is different here is that we don't see the typical pattern of building the number of tickets. But there is a clever trick. Look at this dictionary:

Math	English
550-x	# children's tickets
x	# adult tickets

If we know the total number of tickets sold, and the adult tickets are x, then the children's tickets must be the total − x. (We could have gone the other way too. The number of children's tickets could have been x and the adult tickets could have been 550-x.)

Now, we have a complete dictionary, so we multiply the number of each ticket by their value to create this equation.

$$12x + 7(550 − x) = 4750$$

o) Solve. Remember to give the number of both adult and children's tickets.

p)   Work this on your own.

A children's theater sold 672 tickets to a play. Children's tickets cost $5 and adult tickets cost $7. If the children's theater made $3674, how many of each type of ticket did they sell?

Here, we continue working with word problems. The concepts are the same with just one twist. Our first type of problem will be mixtures.

A trail mix consists of raisins and oats. You need 6 lbs. Raisins cost $2 per pound and oats cost $7 per pound. If the mix will cost $5 per pound, find how many pounds of oats and raisins that you need.

First, our dictionary.

Math	English
	lbs of Oats
	lbs of Raisins

We need the pounds of each. But we aren't given a description of how one builds on the other. Instead, we have the total number of pounds. So:

Math	English
6-x	lbs of Oats
x	lbs of Raisins

This is the pounds of each, but they also have a value, which we account for when we build the equation.

$$2x + 7(6 - x) =$$

But what goes on the right side of the equation. We need 6 pounds, so maybe we put a 6.

$$2x + 7(\cancel{6} - x) = 6$$

This doesn't work. The reason why is that what is on the left must be telling the same story as what is on the right. This equation has the total cost of the mix on the left. On the right it has the pounds of the mix. That doesn't match.

Okay, suppose we put the cost of the mix on the right.

$$2x + 7(\cancel{6 - x}) = 5$$

That doesn't work either. The left is the total cost of the mix and the right is the cost per pound of mix. The answer is we need the total cost of the mix on the right.

$$2x + 7(6 - x) = 6(5)$$

Multiplying the total pounds by the cost per pound gives us the total cost.

a)   Solve the equation. You will get a decimal for x. It is money, so go to two decimal places. And don't forget you need to give the pounds of raisins and oats. (Use the dictionary for help.)

$$2x + 7(6 - x) = 30$$

b)   Work another.

A trail mix consists of raisins and oats. You need 8 lbs. Raisins cost $3 per pound and oats cost $6 per pound. If the mix will cost $4 per pound, find how many pounds of oats and raisins that you need.

In this next problem, the per pound price involves a decimal.

c)   You need 8 pounds of a mix consisting of a cereal and chocolate chips. The cereal costs $5 per pound. The chocolate chips cost $6. The mix will cost $5.50 per pound. Find the number of pounds you need of cereal and the pounds you need of chocolate chips.

This section concludes with financial problems. They follow the same ideas as the mixture problems.

You have $10,000 invested in two funds. One fund makes 5% interest. The other fund makes 7%. How much should you invest in each fund if you want to make 5.5% on the total investment.

Let's begin with the dictionary.

Math	English
x	5% fund
10000-x	7% fund

Notice how I build the dictionary using the total amount. To make the equation, we need to change the percentages to decimals and we get:

$$.05x + .07(10000 - x) = 10000(.055)$$

I've got the total value of the funds on the left, so I put the total value on the right.

$$.05x + .07(10000 - x) = 550$$

I would do the decimal trick and multiply every group by 100.

$$5x + 7(10000 - x) = 55000$$

d) Finish solving, and don't forget to give the amount of money in both the 5% fund and the 7% fund.

Try one on your own.

e) You have $15,000 invested in two funds. One fund makes 3% interest. The other fund makes 8%. How much should you invest in each fund if you want to make 4.5% on the total investment.

Word problems are already difficult, but varying what needs to go on the right side is really challenging. Remember, the key is that the left side and the right side of the equation must be telling the same story.

Algebra I

Active Lesson: 3.4a

In this activity, we want to work with the properties of triangles. First, answer this question:

a) What must all of the angles in a triangle add together to total?

That is all the information that we need to answer basic problems about the sum of the angles in triangles. Here's an example:

Two of the angles in a triangle measure 52° and 87°, find the measure of the missing angle.

Here's my dictionary.

Math	English
x	Missing Angle

And knowing that the angles must add to be 180°, we can create this equation:

$$52 + 87 + x = 180$$

b) Simplify this equation and solve for the measure of the missing angle.

Try one on your own.

c) Two of the angles in a triangle measure 81° and 37°, find the measure of the missing angle.

Some problems will hide the measure of one of the angles inside the definition of a right triangle.

One angle of a right triangle has a measure of 35°. Find the measure of the missing angle.

By definition, a right triangle has one angle which is 90°, so we still know two of the three angles. And our equation is:

$$35 + 90 + x = 180$$

d) Finish solving for the missing angle.

In this next problem, two angles are missing, but we have enough information to build an algebraic equation.

In a right triangle, the measure of one angle is 32° more than the smallest angle.

Here is my dictionary:

Math	English
x	Smallest Angle
x+32	Next Angle

e) Knowing that the triangle is a right triangle, create a formula to solve for the missing angles. In your answer, you must give the value of both missing angles. (So, use your dictionary.)

The concept is the same with the perimeter of a triangle. The perimeter is like a fence. We add up the length of the fence required to cover the outside of the entire shape.

The perimeter of a triangle is 74 meters. One side has a length of 12 meters. Another side has a length of 42 meters. Find the length of the missing side.

Here is the dictionary:

Math	English
x	Missing Side

f)   The only difference here is that the sides of the triangle add up to be the perimeter. Make an equation and find the length of the missing side.

Try one on your own.

g)   The perimeter of a triangle is 102 feet. One side has a length of 49 feet. Another side has a length of 31 feet. Find the length of the missing side.

Finally, we want to work with the area of a triangle. The formula for the area of a triangle is:

$$A = \frac{1}{2} \cdot b \cdot h$$

The problems will give us the area and either the base or the height and we must use our algebra skills to find what is missing.

The area of a triangle is $35 \ m^2$. If the height is $5 \ meters$, what is the length of the base?

Math	English
b	Missing base

So, we plug in the values we've been given.

$$35 = \frac{1}{2}b(5)$$

Remember, the order we multiply doesn't matter, so I'm going to move the b to the end of the equation.

$$35 = \frac{1}{2}(5)b$$

There are many ways you could solve this now. I'm going to multiply the 5 by one-half and then divide.

$$35 = 2.5b$$

$$87.5 = b$$

Try these:

h)   The area of a triangle is $90\ ft^2$. If the height is $12\ feet$, what is the length of the base?

i)   The area of a triangle is $250\ m^2$. If the height is $16\ meters$, what is the length of the base?

Algebra I

Active Lesson: 3.4b

Here, we are going to use the same principles we have been using to solve problems involving the Pythagorean Theorem.

$$a^2 + b^2 = c^2$$

The idea behind the Pythagorean Theorem is this. In right triangles, if you add up the square of each of the two sides (called legs), it always equals the square of the side across from the right angle. (This side is called the hypotenuse.)

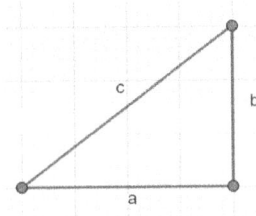

Problems of this type will give us two of the sides and we use algebra to solve for the missing side. For example:

A right triangle has sides of 3 and 4, find the length of the hypotenuse.

So:

$$3^2 + 4^2 = c^2$$

Now, we clean up the equation and solve for c. There is only one new step:

$$9 + 16 = c^2$$

$$25 = c^2$$

We don't want c^2, we just want c. But the opposite of a square is a square-root. If we do something to one side, we must do it to the other.

$$\sqrt{25} = \sqrt{c^2}$$

$$5 = c$$

Try one on your own.

a) A right triangle has sides of 5 and 12, find the length of the hypotenuse.

One variation you will see involves being given the hypotenuse and one leg. Solving the equation takes one more step.

A right triangle has a hypotenuse of 17 and a leg of 15. Find the missing leg.

$$a^2 + 15^2 = 17^2$$

$$a^2 + 225 = 289$$

$$a^2 = 64$$

b) Finish solving for a.

Work this problem.

c) A right triangle has a hypotenuse of 13 and a leg of 5. Find the missing leg.

The final type of problem which we will see in this section is when the two legs of the triangle are the same.

The hypotenuse of a right triangle is 15. Both legs are the same length. Find the length of the legs.

Here is what the equation would look like:

$$x^2 + x^2 = 15^2$$

x^2 are like terms, so they can be combined.

$$2x^2 = 225$$

Now divide by 2 to get x^2 alone.

$$x^2 = 112.5$$

Square rooting both sides gets us a decimal.

$$x \approx 10.6066$$

Round to one decimal place. This will get you the length of both legs.

$$x = 10.6$$

Try one.

d) The hypotenuse of a right triangle is 12. Both legs are the same length.

202

This next section follows all the same principles as the previous section, but this time with rectangles. First, let's look at the perimeter of a rectangle. Remember, perimeter is like adding up fencing around the outside of the shape. Since the lengths are the same, and the widths are the same, the perimeter could be summed up as:

$$P = 2L + 2W$$

Now, suppose we knew the perimeter and either the length or the width. For example:

The perimeter of a rectangle is 120 meters. The width is 20 meters. Find the length.

$$120 = 2L + 2(20)$$

a) Finish solving this for length.

Try these on your own.

b) The perimeter of a rectangle is 180 feet. The width is 65 feet. Find the length.

c) The perimeter of a rectangle is 240 yards. The length is 85 yards. Find the width.

Next, we don't know either the length or the width. We will need to build one from the other.

The perimeter of a rectangle is 200 meters. The length is 5 meters less than the width. Find the width.

Here's the dictionary:

Math	English
x-5	Length
x	Width

This is what the equation would look like:

$$200 = 2(x - 5) + 2x$$

Notice that you would have a lot of clean up to do to solve for x.

$$200 = 2x - 10 + 2x$$

$$200 = 4x - 10$$

$$210 = 4x$$

$$52.5 = x$$

And, you must give both the length and the width. So, using the dictionary:

Math	English
(52.5-5)=47.5	Length
52.5	Width

Work these on your own.

d) The perimeter of a rectangle is 250 meters. The length is 15 meters less than the width. Find the width.

e) The length of a rectangle is five more than twice the width. The perimeter is 70 feet. Find the width.

Finally, we will solve area problems involving rectangles. The area of a rectangle is the length times the width.

$$A = L \cdot W$$

The ideas are the same.

If the area of a rectangle is $100 \ meters^2$, and the length is 20 meters. Find the width.

f) Set up an equation and solve.

Work this problem on your own.

g) If the area of a rectangle is $250 \ cm^2$, and the width is $12.5 \ cm$. Find the width.

Algebra I

Active Lesson: 3.5a

How far did each of the following travel?

    a) Car one drives 55 mph for 3 hours.

    b) Car two drives 33 mph for 5 hours.

    c) Train one moves at 70 mph for 5.5 hours.

    d) Train two moves at 55 mph for 7 hours.

e) What was true about the distance that car one and car two travelled?

f) What was true about the distance that train one and train two travelled?

We are going to begin some difficult word problems, but the tools we've practiced should help to make it manageable. This section is based on the formula:

$$Distance = Rate \cdot Time$$

How fast you are going, multiplied by how long you go that fast, equals your distance.

Your brother left home and drove for 5 hours. Later, you realized that he forgot something and you went after him. You caught up after driving for 3 hours. Your rate was 12 mph faster than your brothers. What were the rates of the two cars?

To answer this question, we need a dictionary that is based on the formula: $D = R \cdot T$.

	Rate	Time	Distance
Brother's Car			
Your Car			

Let's fill in what we know. We know the times.

	Rate	Time	Distance
Brother's Car		5	
Your Car		3	

We don't know the rates, but we can use the ideas from the last few sections to build them.

	Rate	Time	Distance
Brother's Car	r	5	
Your Car	r+12	3	

Because $D = R \cdot T$, we can always fill in the distance column. Multiply the rate column by the time column gets us:

	Rate	Time	Distance
Brother's Car	r	5	5r
Your Car	r+12	3	3(r+12)

Next, we need a formula. Here's what we know. Your car caught up to your brother's car. So, you travelled the same distance. It may not look like it, but the values in the last column are distances. And, we know each of those distances are the same.

$$5r = 3(r + 12)$$

g) Remember, we always need the left side and the right side to be saying the same thing. The left side is the total distance your brother travelled. The right side is the total distance you travelled. And, we know they are equal. So, we have our formula. Finish solving the equation. (You will need to get all the r's back together in the same room.)

h) Give the rates of both your brother's car and your car. Use the dictionary to help.

	Rate
Brother's Car	r
Your Car	r+12

Try one on your own.

Two trains left from the same station. The first train travelled for 7 hours. The second train was the express train and reached the same destination after 5.5 hours. The rate of the express train was 15 mph faster than the regular train.

i) Complete the dictionary. Remember, the distance column is always the rate column multiplied by the time column.

	Rate	Time	Distance
Standard Train			
Express Train			

j) To make your equation, use the fact that the express train travelled the same distance as the standard train.

k) Use your dictionary to give both the rate of the standard train and the express train.

What makes these problems difficult is the variations. Let's look through another.

Two trains start 150 miles apart before travelling to meet each other. The first train travels for 1 hour. The second train travels for 2 hours. The second train is 15 mph faster than the first. How fast are each of the trains moving?

Let's begin with the dictionary:

	Rate	Time	Distance
First Train	r	1	
Second Train	r+15	2	

And, as before, the distances are the rates multiplied by the times.

	Rate	Time	Distance
First Train	r	1	1r
Second Train	r+15	2	2(r+15)

However, making our formula will be different this time. We need the left side and the right side to be telling the same story, but the two trains didn't travel the same distance. However, we know the total distance which they travelled. That distance was 150 miles. If we *added* the distance of the two trains than they must add to 150.

$$1r + 2(r + 15) = 150$$

Here, the left side is the total distance travelled by the trains *and* the right side is the total distance travelled by the trains.

l) Solve the equation for r.

m) Use the dictionary to give both rates.

	Rate
First Train	r
Second Train	r+15

Try one on your own.

Two trains start 200 miles apart before travelling to meet each other. The first train travels for 2 hours. The second train travels for 3 hours. The rate of the second train is 10 mph faster than the first train.

n) What are the rates of the two trains?

	Rate	Time	Distance
First Train			
Second Train			

So far, we've worked with problems looking for the rates. In this final type of problem, we will need to find the time.

Two cars start at the same point. One car drives east. The other car drives west. If one car drives 55 mph and the other car drives 65 mph, how long will it take for the two cars to be 420 miles apart.

As always, we start with the dictionary.

	Rate	Time	Distance
First Car	55	t	
Second Car	65	t	

And, as before, we always know the distance column by multiplying the rate and time.

	Rate	Time	Distance
First Car	55	t	55t
Second Car	65	t	65t

To build our formula, we need the left and right sides to tell the same story. At the end, we know that the cars are 420 miles apart. That is a total distance:

$$55t + 65t = 420$$

The left and right sides are both the total distance which the cars have travelled. Solve the equation for t

o) (time). There is only one answer here, since both cars travelled for the same amount of time.

Work this problem on your own.

p) Two cars start at the same point. One car drives east. The other car drives west. If one car drives 70 mph and the other car drives 65 mph, how long will it take for the two cars to be 472.5 miles apart.

Algebra I

Active Lesson: 3.5b

Find the following distances:

    a)   A car travels 60 miles per hour for 3 hours.

    b)   A car travels 60 miles per hour for 30 minutes.

    c)   A car travels 60 miles per hour from 1:00 pm to 3:00 pm.

    d)   A car travels 60 miles per hour from 1:30 pm to 3:00 pm.

When solving problems involving $d = r \cdot t$, we need to be sure that the rate and time are in the same units. It can be done as either fractions or decimals. Convert each of the following to both a fraction and a decimal that matches the rate. This will typically be hours. We want part of an hour. So, with 60 minutes in an hour, we can divide. I'll help with the first one.

15 minutes in hours

$$\frac{15}{60} = \frac{1}{4} = .25 \ hours$$

e)   20 minutes in hours

f)   45 minutes in hours

g) 30 minutes in hours

Other than getting the rate and time in the same units, the ideas here are identical to the last activity.

You and your brother are going to school. He walks and it takes him 30 minutes. You ride your bike and it takes 15 minutes. With your bike, you can travel 5 mph faster. Find the rates of you and your brother.

The bike is 5 miles per hour faster, so to make this work, the entire problem should be in hours. First, we build our dictionary.

	Rate	Time	Distance
Your Brother	r	0.5	
You	r+5	0.25	

Next, we fill in the distance column:

	Rate	Time	Distance
Your Brother	r	0.5	.5r
You	r+5	0.25	.25(r+5)

To make the equation, we notice that you and your brother both travelled the same distance. So, we will set the two distances equal to each other.

$$.5r = .25(r + 5)$$

h) Solve for r. (You can work with the decimals or multiply each group by 100. There are only two groups; one on the left and one on the right.)

i) Use the dictionary to give both the rate of your brother and your rate.

	Rate
Your Brother	r
You	r+5

Try this problem.

j) You and your brother are going to school. He walks and it takes him 45 minutes. You ride your bike and it takes 30 minutes. With your bike, you can travel 6 mph faster. Find the rates of you and your brother.

Let's look at a second type of problem.

A biathlon involves swimming and biking. The winner swam from 1:00 p.m. until 2:30 p.m. Then biked from 2:15 p.m. until 4:00 p.m. The rate of biking was three times faster than the swimming. If the entire race covered 67.5 miles, find the rate of the biking and the swimming.

First, the dictionary.

	Rate	Time	Distance
Swimming	r	1.5	
Biking	3r	1.75	

As usual, the distances are the rates multiplied by the times.

	Rate	Time	Distance
Swimming	r	1.5	1.5r
Biking	3r	1.75	1.75(3r)=5.25r

We know the total distance of the race, so to make a matching equation, we get:

$$1.5r + 5.25r = 67.5$$

k) Solve for r.

l) Give the rate for both the swimming and the biking.

Try a problem on your own.

m) A car is stuck in slow moving traffic from 2:00 p.m. until 2:45 p.m. Afterward they drive 2:45 p.m. to 4:00 p.m. When they leave the slow traffic, their rate is four times as fast. If the total journey was 86.25 miles, find the rate in slow moving traffic and the rate afterward.

	Rate	Time	Distance
Slow			
Faster Traffic			

Algebra I

Active Lesson: 3.6

We are going to continue our translation work with another section of word problems. Unlike what we've seen before, however, these problems will involve inequalities. To begin, translate each of the following statements into one of these four inequality signs: $>, \geq, <, or \leq$.

    a) No more than.

    b) Greater than.

    c) Fewer than.

    d) The maximum allowed is...

    e) Must exceed.

    f) At least.

Let's begin translating.

During a school field trip, the class encounters an elevator which has a maximum weight capacity of 2,000 lbs. If the average weight of a student in the class is 110 lbs, how many students can get on the elevator at one time?

Here is our dictionary:

Math	English
x	Number of Students

Let's translate between English and math language.

Maximum weight of 2000: You are allowed to have 2000 lbs, but no more.

$$\leq 2000.$$

Average weight of 110: This isn't a clear translation, but it is 110 lbs per student. So:

$$110x$$

And our equation becomes:

$$110x \leq 2000$$

Solving this seems straightforward, but there is a twist:

$$x \leq 18.18$$

We needed to know the maximum number of students allowed on the elevator and we can't have part of a student. So, we would round down and no more than 18 students would be allowed on.

$$x \leq 18$$

Try a problem on your own.

g) The teacher on the field trip wants to buy the students a souvenir. She has $25. The souvenirs cost $1.35 each. What is the maximum number of souvenirs she can buy?

Let's try another.

A car salesman's bills are $4,200. He makes a flat salary of $550 and then $660 dollars per car sold. What is the least number of cars that he can sell in a month in order to pay his bills?

Math	English
x	# of Cars Sold

Here's the translation:

A flat salary of $550 and then $660 per car sold. The flat salary doesn't change. Then, 660 per car sold:

$$550 + 660x$$

He must make at least $4,200.

$$\geq 4200$$

So:

$$550 + 660x \geq 4200$$

h) Solve for x. (You will get a decimal. This time, we need "at least," so we will need to round up to the next highest whole number.)

Work this problem on your own.

i) A television salesman needs to make $2,100 this month. The base salary is $475 plus $125 per television sold. What is the fewest number of televisions he can sell?

There can be lots of variations to this type of problem, however let's look at another.

A teenager has started a lawnmowing company. She charges $65 per lawn and has expenses of $350. What is the fewest number of lawns she can mow in order to make a profit of at least $1500?

Math	English
x	# of lawns mowed

This is a story about her profit. Profit is income minus expenses.

Translation:

$65 per lawn:

$$65x$$

Expenses of $350:   (This is negative because the story is about profit.)

$$-350$$

A profit of $1500:

$$\geq 1500$$

Putting this all together, we subtract the expenses from the income and get:

$$65x - 350 \geq 1500$$

j) Solve for x. (You will need to round up again.)

Work this problem.

k)   A babysitter charges $26 per hour. She has expenses of $1250. How many hours must she work in order to make a profit of at least $1300?

Algebra I

Active Lesson: 4.1a

In algebra, one of our major ideas is graphing. The basis of the concept is like a treasure hunt.

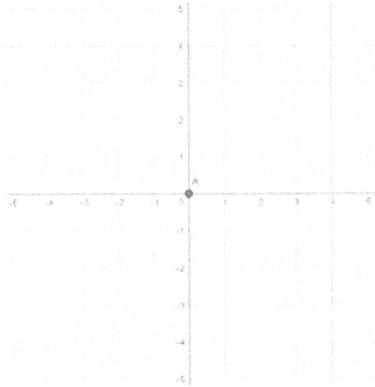

Imagine a treasure has been buried in a field. To find it, the field has been divided into a grid.

You start at the middle (called the origin). The instructions tell you where to go.

The first instruction tells you whether to go right or left. (Positive is right. Negative is left.) The second tells you whether to go up or down. (Positive is up. Negative is down.)

Here is an example: $(-2, 3)$

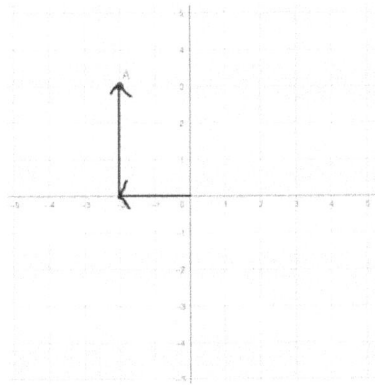

The first instruction was -2. So, I went left 2.

The second instruction was 3. So, I went up 3.

Try some on your own. Start at the origin (0,0) for each.

a) $(3, -2)$ b) $(-1, 4)$ c) $(-2, -4)$

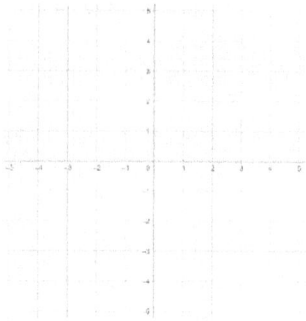

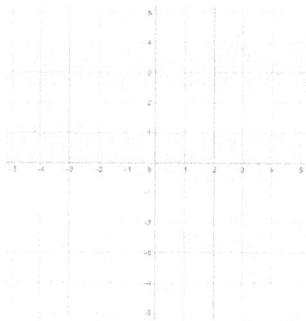

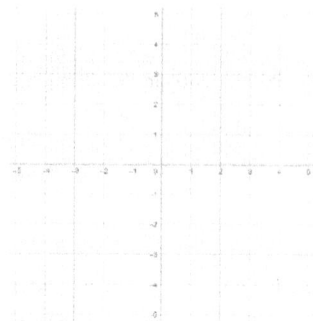

This system divides the field into four boxes called Quadrants. The boxes go counterclockwise.

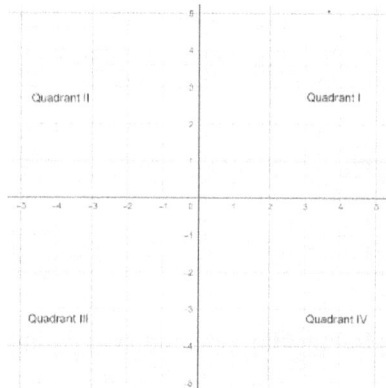

The point $(2, -3)$ would be in quadrant IV.

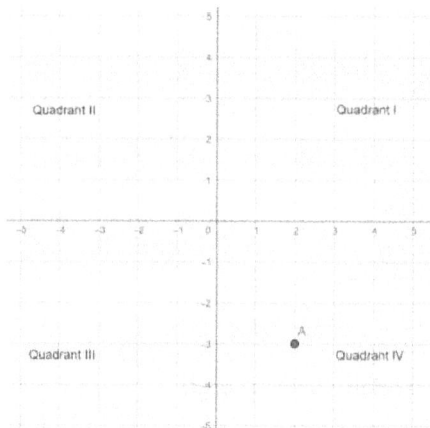

Graph the following points *and* state which quadrant they would be in:

d) $(-2,3)$ e) $(1,3)$ f) $(-1,-3)$

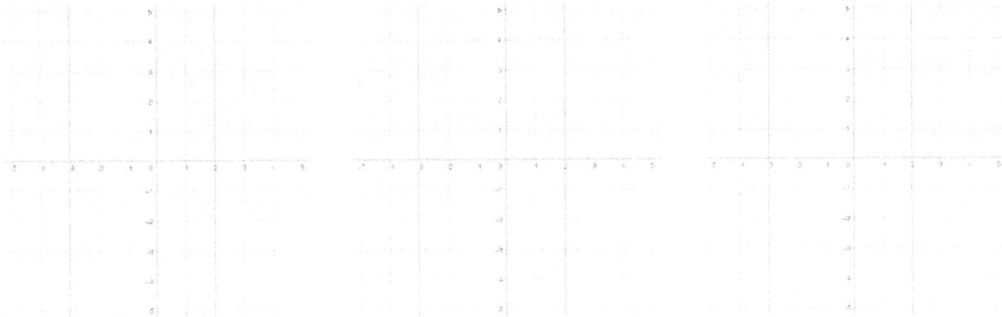

A point isn't always in a quadrant. They can also be on one of the two major lines.

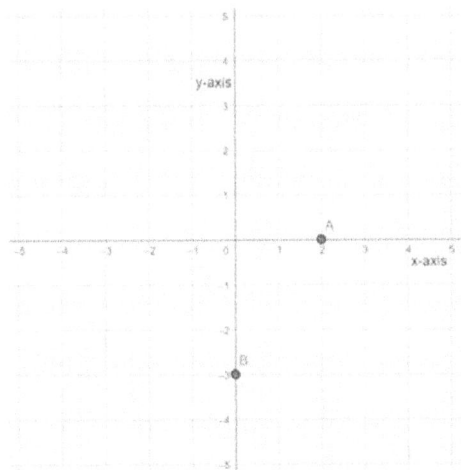

The right and left line is called the x-axis.

The up and down line is called the y-axis.

Point A is on the x-axis. Point B is on the y-axis.

Which axis would the following points be on:

g) $(3,0)$ h) $(0,4)$ i) $(-1,0)$

In this system, a point is called an ordered pair. The first instruction, right or left, is called the x-coordinate. The second instruction, up or down, is called the y-coordinate. So: (x,y).

j) For the following points, state the x and y coordinates:

$(-2, 3)$

x-coordinate:

y-coordinate:

k) $(-1, -3)$

x-coordinate:

y-coordinate:

l) $(8, 12)$

x-coordinate:

y-coordinate:

m) $(0, 0)$

x-coordinate:

y-coordinate:

n) The final point has a special name. What was its name?

Algebra I

Active Lesson: 4.1b

Graphing is one of the main concepts of algebra. And it is built on the ideas of ordered pairs and relations. We saw ordered pairs in the last activity. Today we want to look at relations. Here's an example.

$$x + y = 8$$

The equal sign makes this an equation. But because there are two variables, this also means something else. The two variables are in relationship with one another. If you add x and y you get 8. Yet since these are variables, there are lots of ways to do it. I showed you the first one. Finish the table.

a)

x	y
3	5
5	
2	
-1	
-4	

There are more than this. In fact, there are an infinite number of combinations. Each working combination is called a solution to that equation, and they create an ordered pair. I've plotted the first ordered pair $(3, 5)$ on the graph below. Add the rest of the ordered pairs from the table.

b)

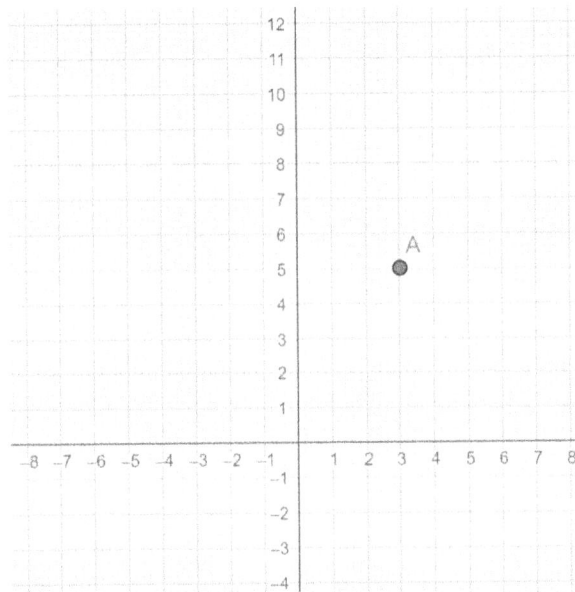

Any relationship (just called a relation in math) between an x and a y makes a line. Connect the points from the table which you graphed. (If your points are correct than you should have a line.) But there are infinite possible points and so we extend the line in both directions and add arrows on the ends. (Extend your line and add the arrows.)

Here is another relation. This one is a little bit more complicated. You may need your algebra skills.

$$2x + y = 8$$

I'm going to start with $x = 3$ and find the y which is his partner.

$$2(3) + y = 8$$

$$6 + y = 8$$

$$y = 2$$

I've done the first two ordered pairs. Finish the table.

c)

x	y
3	2
2	4
5	
1	
-1	

Graph the ordered pairs and make a line.

d)

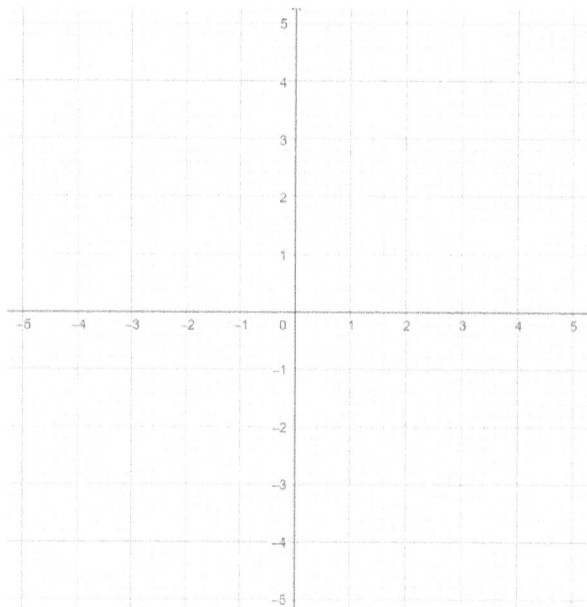

Here's another. The relation is presented differently; however, it is still the same idea.

$$y = 4x - 3$$

e)

x	y
0	
-1	
2	

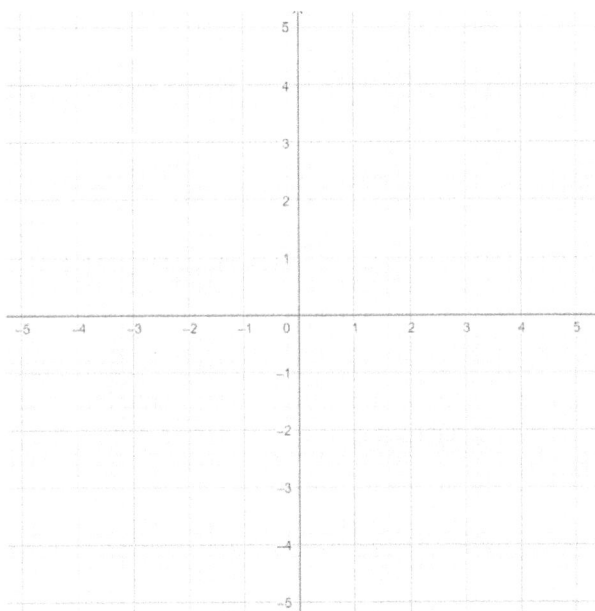

Because the line goes on forever, every value of x has a partner value of y. So, you could actually pick any value of x you wanted and then find the partner y. The problem is that some of them turn out to be fractions which are difficult to graph. If you are picking your own points, make choices which cause the math to be easy. (For instance, I always use 0.) Three points are all that is required to graph a line.

$$y = 3x - 4$$

Pick your own points and graph the line:

f)

x	y

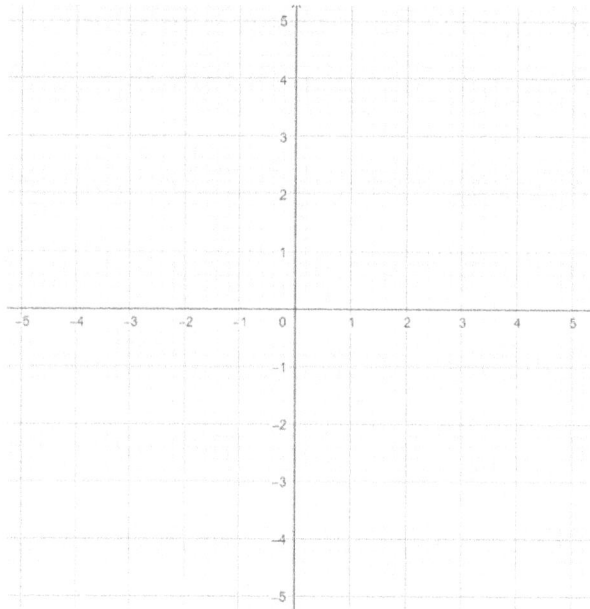

Finally, any ordered pair which makes the relation true is called a solution. If you graphed the line, solutions are ordered pairs which are on the line.

Look at this relation:

$$3x + 2y = 12$$

g) Circle any of the following ordered pairs which are solutions:

    a) $(0, 6)$
    b) $(-2, 8)$
    c) $(-4, 12)$
    d) $(6, -3)$
    e) $(10, 9)$

Below is the graph of a line:

$$y = 3x + 1$$

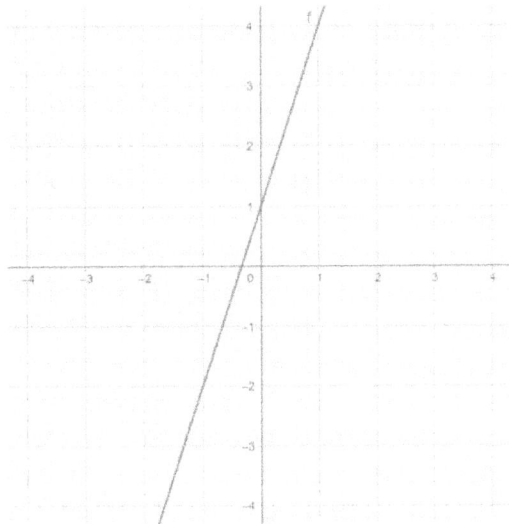

a)   An ordered pair is a solution if it is on the line. Circle any of the following which are a solution to the line: $y = 3x + 1$.

    a) $(1, 4)$
    b) $(-1, -1)$
    c) $(-1, -2)$
    d) $(0, 2)$

To graph a line, you need three points. Any values of x will do, but picking easy values helps with the work. With fractions, let x be the same as the denominator.

$$y = \frac{1}{4}x - 3$$

$$y = \frac{1}{4}(4) - 3$$

$$y = 1 - 3$$

$$y = -2$$

When your x value is the same as the denominator, it will cancel. Finish the table below and then graph the line.

$$y = \frac{1}{4}x - 3$$

b)

x	y
4	-2
0	
-4	

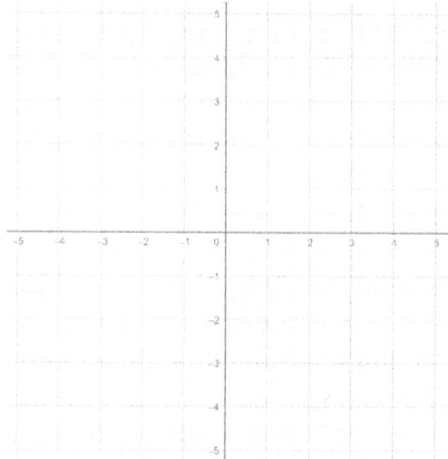

Try another: $y = \frac{3}{5}x - 2$

c)

x	y
5	
0	
-5	

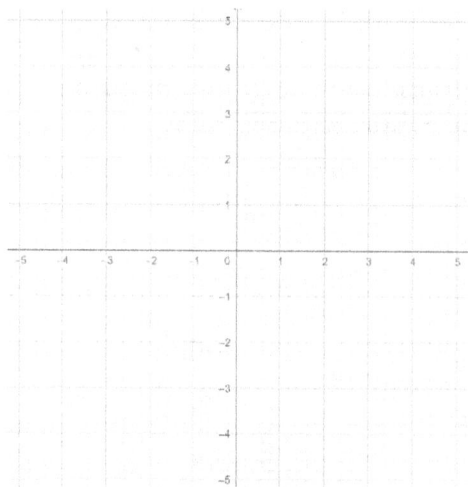

Algebra I

Active Lesson: 4.2b

In this activity, we are going to graph very strange lines. Look at this one:

$$y = 3$$

It says that y is always 3. So, no matter what x values we choose, y will be 3. Complete the table and graph the line. (Don't over think it.)

a)

x	y
1	
0	
-1	

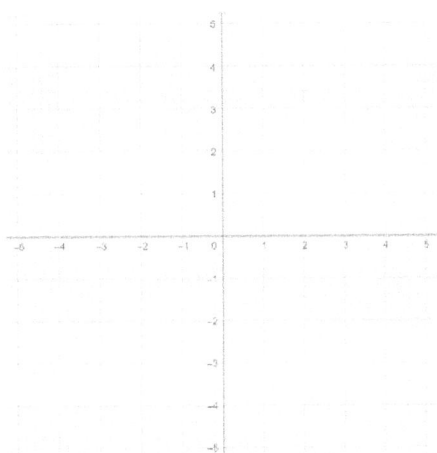

Once you see what will happen, you could have done this without the table. Graph $y = -1$. The y's will always just be -1.

b)

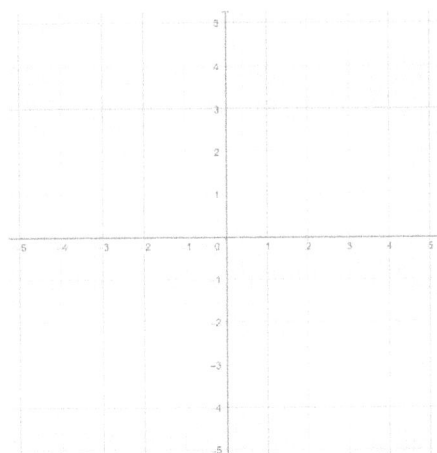

Now, let's try one with just x.

$$x = -2$$

Our table is backward, but it works the same. The values of x will always be -2.

c)

x	y
	-1
	0
	1

Graph the line.

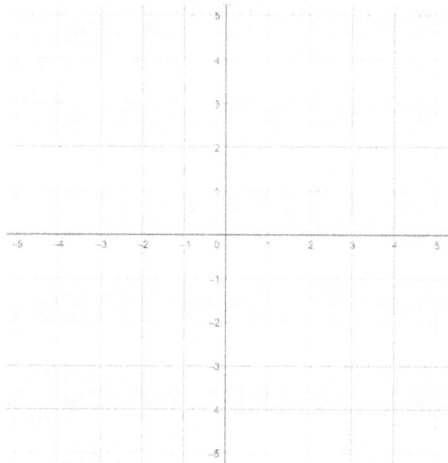

Again, the table is unnecessary. Graph the line $x = 1$. (x is always 1.)

d)

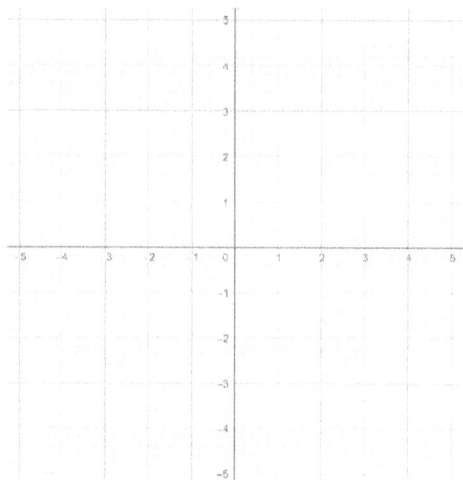

e) If you see the pattern, you know that lines involving $y =$ are always_____.

(Circle the correct answer.)

     a) Horizontal
     b) Vertical

f) Similarly, you know that lines involving x= are always_____. (Circle the correct answer.)

     a) Horizontal
     b) Vertical

It is often very important to know where a line hits the x-axis or y-axis. Where a line hits the x-axis is called the x-intercept. Where a line hits the y-axis is called the y-intercept. Here is the graph of the line $y = 2x + 2$.

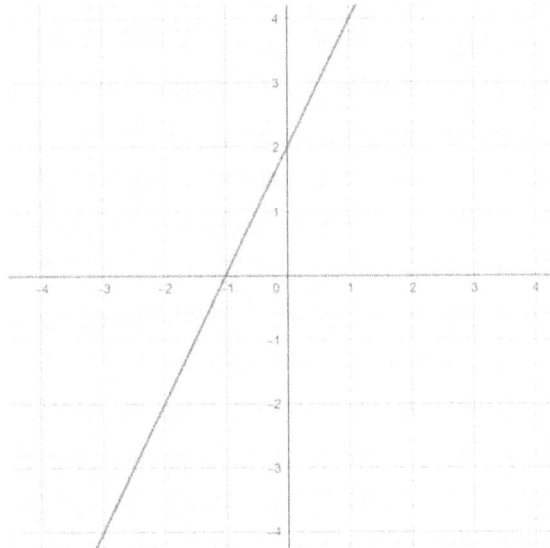

a) Give the value of the intercepts. (They must be given as ordered pairs, not just a single number.)

x-intercept: (__, 0)

y-intercept: (0, __)

The x-intercept always occurs when the y-coordinate is zero. The y-intercept always occurs when the x-coordinate is zero.

Try another. Don't forget to give the intercepts as ordered pairs.

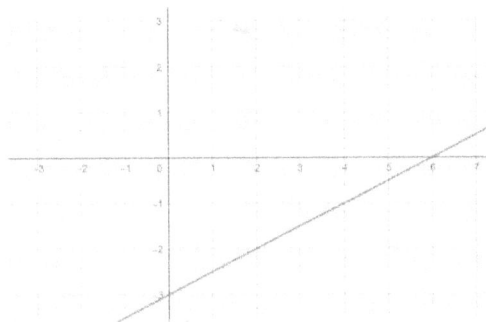

b) x-intercept:

y-intercept:

We have been taking the intercepts off the graph. However, we can find them by entering zeros into the equation. Since the x-intercept occurs when $y = 0$, we can enter a 0 into the equation for y. Likewise, the y-intercept occurs when $x = 0$, and so we can enter a 0 in for x. (Normally, if we are graphing a line, we pick three points, but if you have both the intercepts that is sufficient.)

Find the intercepts for the line $2y + 3x = 6$.

c)

	x	y
x-intercept		0
y-intercept	0	

Plot the intercepts and graph the line.

d)

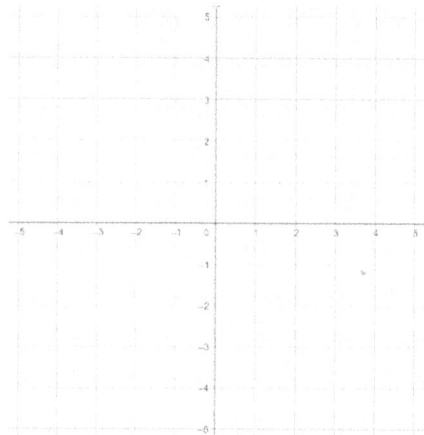

Graph one more line by using the intercepts.

$$5x - 2y = 10$$

e)

	x	y
x-intercept		0
y-intercept	0	

Plot the intercepts and graph the line.

f)

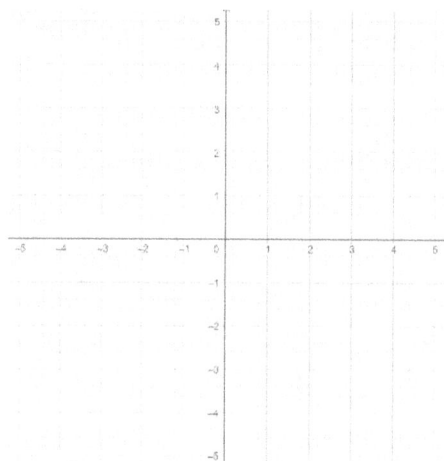

Below is the graph of the line: $y = \frac{2}{3}x - 2$.

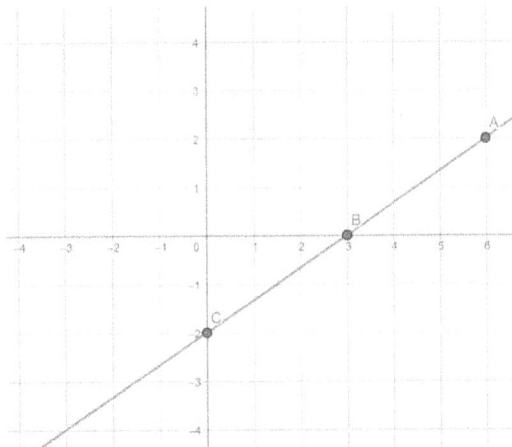

Pretend the grid lines are city block and you are standing at point C. You are going to walk up and then to the right to get to point B.

a) How many blocks up:

   How many blocks to the right:

   Now you are at point B. You are going to walk up and then to the right to get to point A.

b) How many blocks up:

   How many blocks to the right:

c) Here is the formula again: $y = \frac{2}{3}x - 2$. Something in the formula matches with the blocks that you walked between points, what is it?

Up and down is the y direction. Up is positive. Down is negative. Right and left is the x direction. Right is positive. Left is negative.

Start at point A and walk down and then to the left to get to point B. (Remember both of these directions are negative.)

d) How many blocks down:

How many blocks left:

Here's the formula again: $y = \frac{2}{3}x - 2$. The number in front of the x is called the slope. It tells us how to get between any two points on the line. When you went down and left your numbers were negative.

e) Explain why this still gets you the same slope.

Next, look at the graph of the line $y = -\frac{3}{4}x + 1$.

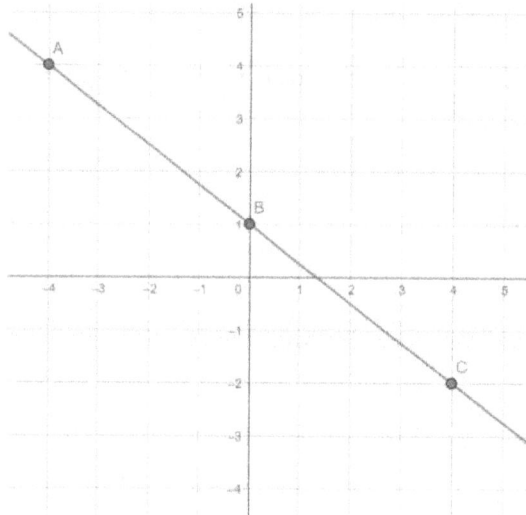

I'm choosing points that are clearly at intersections of the grid. Start at point A and go to point B.

f) How many steps down?   (Remember this is negative.)

g) How many steps to the right? (Remember this is positive.)

h) Is this consistent with the slope in the equation?

This time, start at point B and go back to point A.

i) How many steps up? (Remember this is positive.)

j) How many steps to the left? (Remember this is negative.)

k) Below, write both the slope when we went between point A and point B and the slope when we went between point B and point A.

l) Are they the same mathematically?

Now, start at point A and go to point C.

m) How many steps down:

n) How many steps to the right:

o) What is the slope?

p) Show why this is the same as the slope between points A and B.

Graphing works just like reading. We read from left to right and graphs do the same. A positive slope, goes uphill from left to right. A negative slope goes downhill from left to right.

Find the slope of the following graphs. (Use two points that are clearly at the intersection of the grids.)
Then, put a circle around any lines with slopes that are positive and a box around any lines with slopes
that are negative.

q)

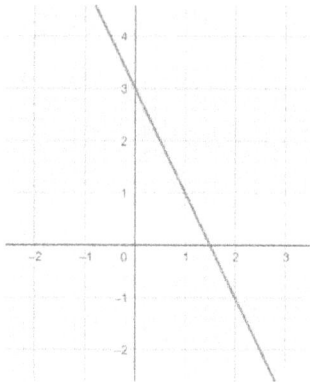

r)

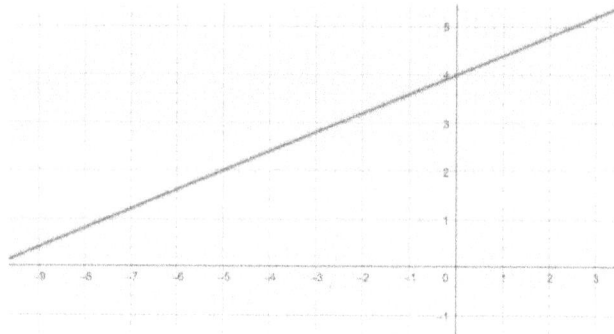

s)

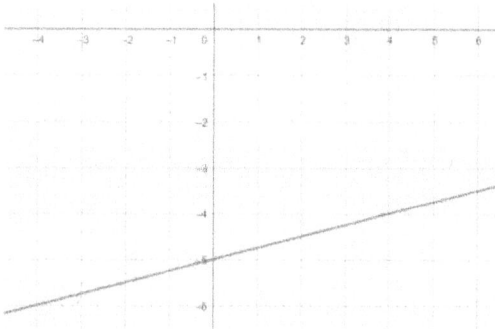

t)

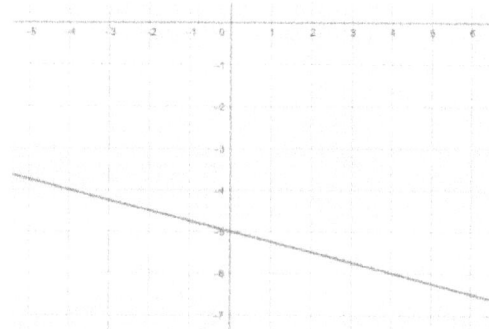

u) Here are four formulas for lines. Circle any that have positive slopes. Put boxes around any that have
negative slopes.

$$y = -\frac{1}{5}x + 2$$

$$y = \frac{3}{2}x - 1$$

$$y = \frac{5}{7}x + 6$$

$$y = -2x + 3$$

Graphing this last line $y = -2x + 3$ seems like a problem. The slope is only -2. However, the solution is easy. Any number can become a fraction by putting it over the number 1. So, it becomes:

$$y = -\frac{2}{1}x + 3$$

v) Graph it below. I've given you a starting point. Use the slope to add two points and then make your line.

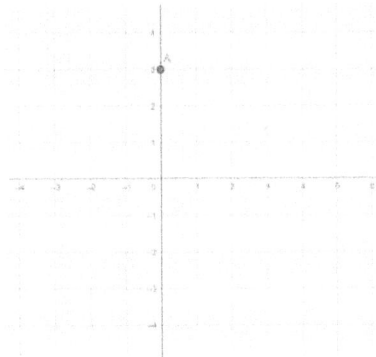

In math language, the slope of a line is called m. And it is easy to find from a graph. But if we didn't have the graph, we can find it from knowing two points on the line, using this formula:

$$m = \frac{y_2 - y_1}{x_2 - x_1}$$

This formula is nothing more than what we've been doing.

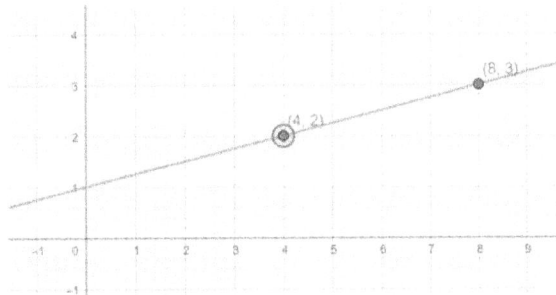

w) From the graph, what is the slope of this line?

x) Using the points, subtract the two y-values.

y) Using the points, subtract the two x-values.

z) Put your answer for the y's over the answer for the x's and you have the slope. (It doesn't matter which point you think of as the first point and which you think of as the second. Just stay consistent. Use the same point as both x_2 and y_2.

Use the formula $m = \dfrac{y_2 - y_1}{x_2 - x_1}$ to find the slope between the following points:

aa) (2, 7) and (4, 10)

bb) (5, 3) and (8, 1)

On this next one, be careful, subtracting a negative number makes two negatives in a row, which makes it addition.

cc) (−4, 5) and (−2, 3)

dd) Below I have graphed the horizontal line $y = 3$. Use the formula and the two points I have marked to find the slope.

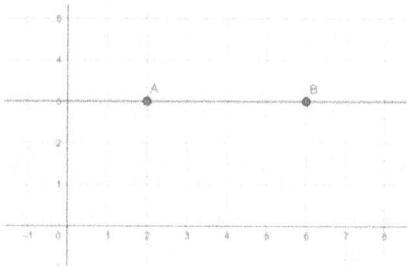

ee) Horizontal lines always have a slope of zero. Why?

ff) Here is a graph of the line $x = 2$. Use the formula and the two points I have marked to find the slope.

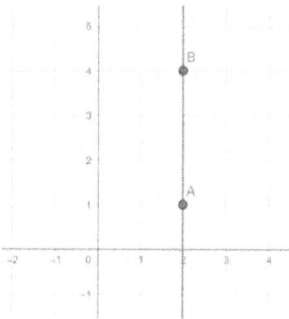

gg) Vertical lines always have an undefined slope. Why?

If we know the slope of a line then we only need one point on that line to graph it. I've given you a point, graph the line if $m = \frac{2}{3}$. (Remember, you need three points to make a line.)

a)

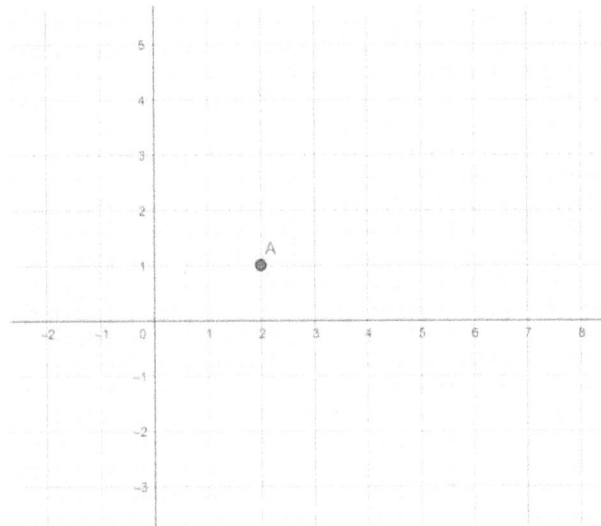

Graph the line given the following information:

$m = -\frac{2}{3}$ and containing the point $(4, 1)$

(To graph a negative slope, give the negative either to the top $\frac{-2}{3}$ or to the bottom $\frac{2}{-3}$. They are the same thing. Don't give the negative to both the top and the bottom. That wouldn't be the same.)

b)

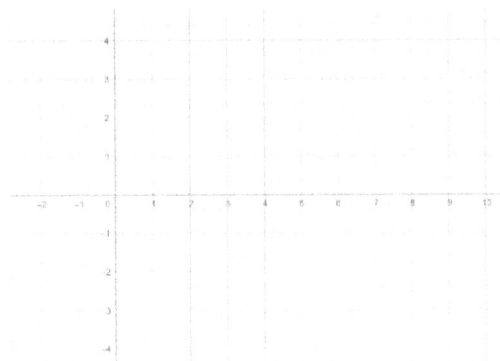

We can find the line with the slope and any point. However, as we will see, the most common point to use is the y-intercept. (Remember, the y-intercept is where the line hits the y-axis.)

Graph the line with the slope $m = \dfrac{1}{2}$

c)

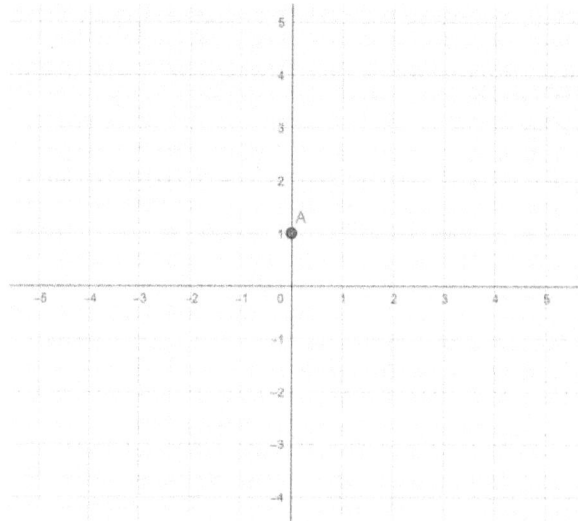

Graph the line with the given information:

$m = -\dfrac{1}{3}$ and y-intercept: $(0, 4)$

d)

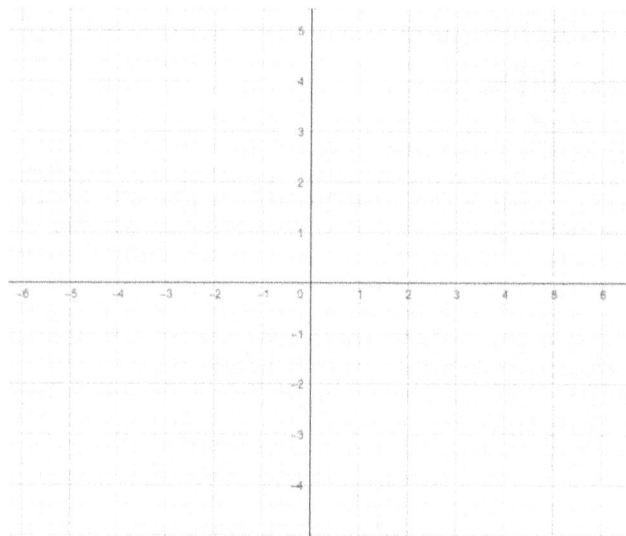

When a slope isn't a fraction, make it a fraction by putting it over 1.

$m = 3$ and y-intercept: $(0, -2)$ (Make this slope $m = \frac{3}{1}$)

e)

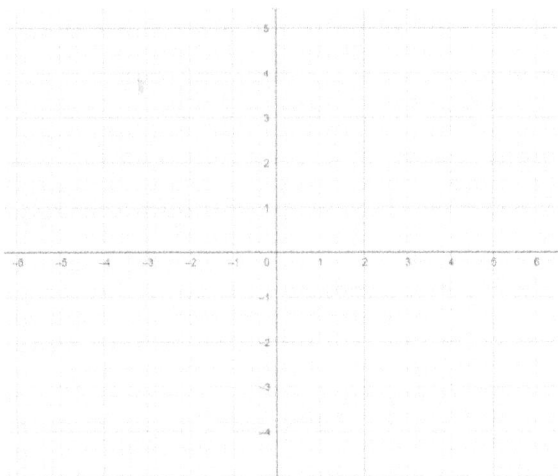

Try one more, but this one has a twist.

$m = 0$ and y-intercept: $(0, 1)$

f)

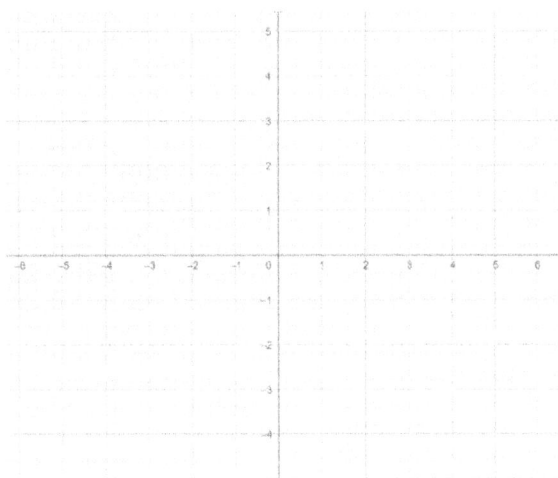

Algebra I

Active Lesson: 4.5a

Find the y-intercept of the following lines:

(Remember, the y-intercept occurs when $x = 0$)

$$y = 3x + 5$$

a)   y-intercept: (0, ___)

$$y = -\frac{2}{5}x + 1$$

b)   y-intercept: (0, ___)

$$y = -\frac{54}{55}x - 3$$

c)   y-intercept: (0, ___)

d)   When the equation for a line is in this form there is a connection with the y-intercept. What is the connection?

This form is called slope-intercept form $y = mx + b$. It tells you everything you need to graph the line. You have the slope (m), the value in front of the x, and the value of the y-intercept, called b.

State the slope and the y-intercept for the following lines:

$$y = -\frac{3}{2}x + 5$$

e) Slope:

y-intercept:

Use the slope and y-intercept to graph the line.

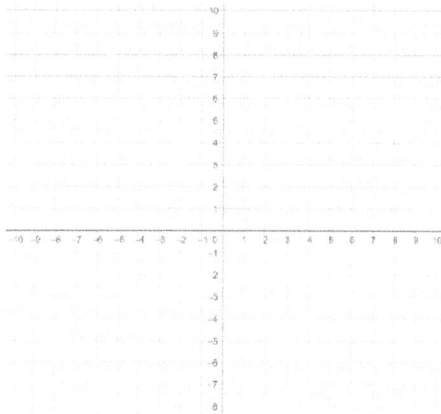

$$y = \frac{4}{5}x - 4$$

f) Slope:

y-intercept:

Use the slope and y-intercept to graph the line.

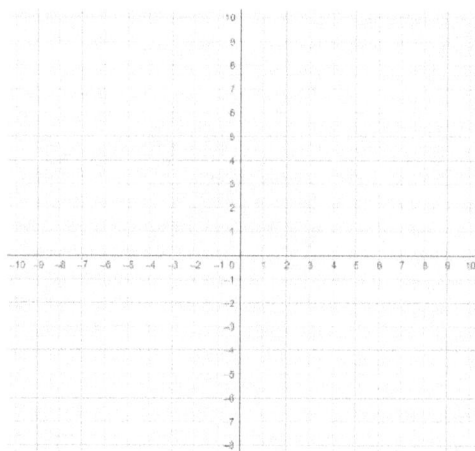

$$y = -\frac{1}{3}x + 7$$

g) Slope:

y-intercept:

Use the slope and y-intercept to graph the line.

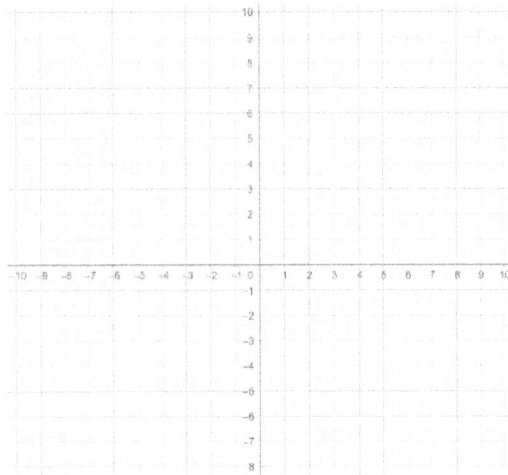

This version of the equation for a line is so helpful that we often want to rearrange an equation into this form. Use your algebra skills to put the following lines in slope-intercept form. (Get y alone on one side.)

h) $y - 2x = 3$ (Add $2x$ to both sides of the equation.)

i) $y - 5x = 8$

j) $y + 3x = 2$

When there is something in front of the y it gets a little more complicated. Let me show you.

$$2y - 4x = 8$$

Send the 4x to the other side.

$$2y = 4x + 8$$

Now, we divide the 2 away from both sides.

$$\frac{2y}{2} = \frac{4x + 8}{2}$$

$$y = \frac{4x + 8}{2}$$

There are two different terms on the right side. The 2 gets divided into both.

$$y = 2x + 4$$

Try one:

k)
$$3y + 6x = 9$$

Sometimes you try to divide but you can't go into both terms.

$$y = \frac{5x + 10}{2}$$

That's okay. It makes the slope.

$$y = \frac{5}{2}x + 5$$

Try one:

l) $$3y + 2x = 12$$

Try another. It is in a bit of a different order, but the ideas are the same. When you have converted it into slope intercept form, graph the line.

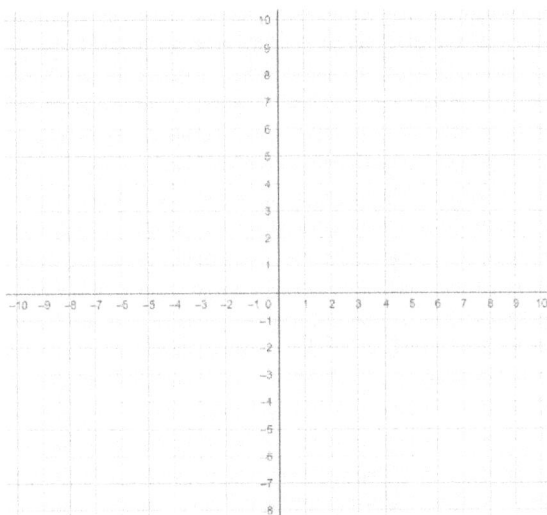

m)
$$4x + 3y = 6$$

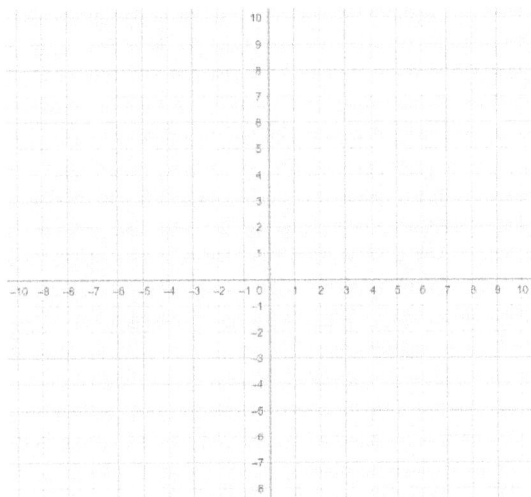

One this final problem, you will need to divide both sides by a -2. But everything works the same.

n)
$$3x - 2y = 4$$

Graph the line:

Algebra I

Active Lesson: 4.5b

As we've looked at lines, we've seen two forms of the equation. The most important form was the slope-intercept, which looks like this:

$$y = -\frac{2}{3}x + 3$$

This form is important because it easily tells us the slope and y-intercept. Give the slope and y-intercept for this line.

a)   Slope: b)   y-intercept:

Use the slope and y-intercept to graph the equation.

c)

The second form of the equation for a line was the standard form. It looks like this:

$$2x + 4y = 12$$

The standard form gives us an easier way to compute the x and y intercepts. Find the intercepts for this line.

d) x-intercept:

e) y-intercept:

We usually use three points to graph a line, but if we have the intercepts, two points are sufficient. Graph the line below, using the intercepts.

f)

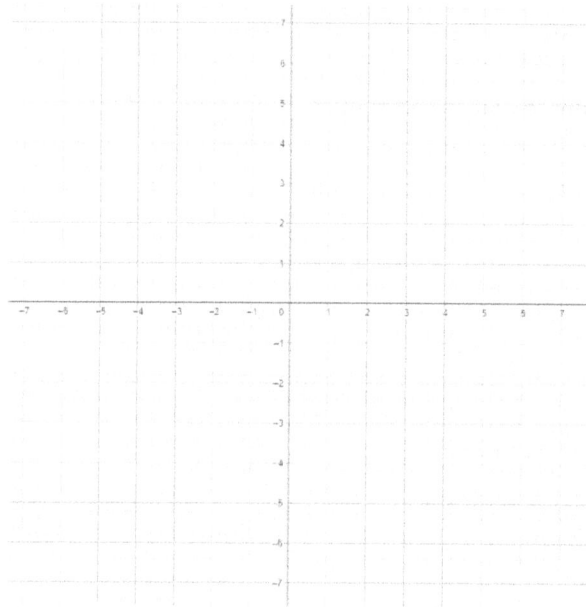

We also encountered equations which made vertical or horizontal lines. Look at the following and indicate if the equation would make a vertical or horizontal line.

g) $x = -2$ vertical or horizontal

h) $y = 5$ vertical or horizontal

Graph each of the lines below.

i) $x = -2$

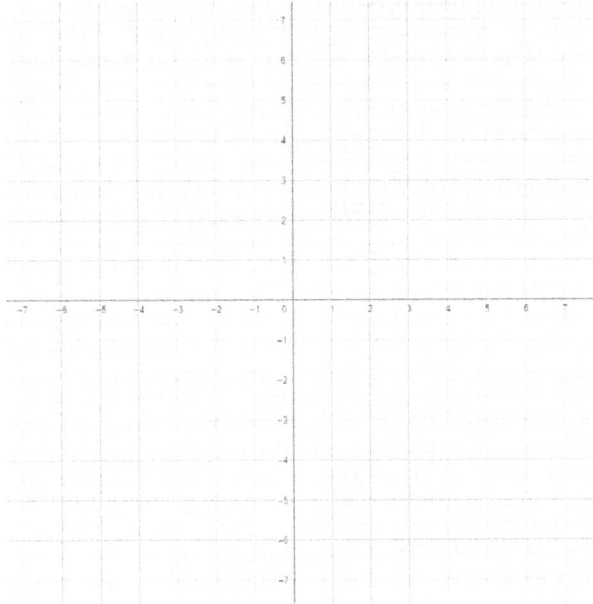

j) $y = 5$

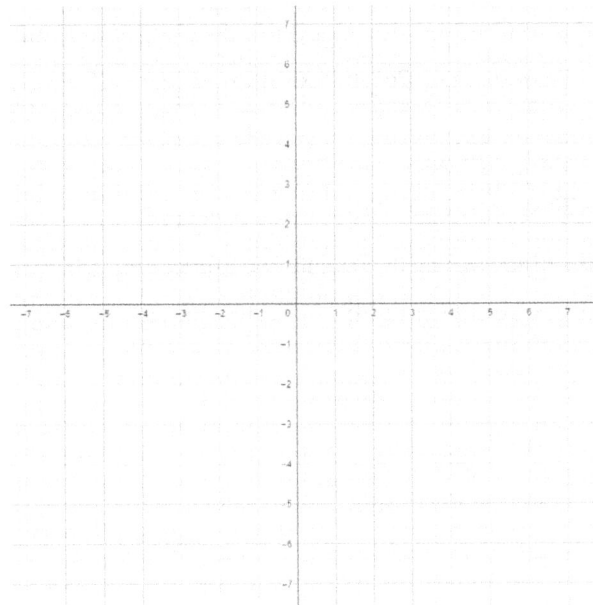

Next, let's look at applications of lines and answer some basic questions.

The equation below estimates the age of a toddler (in months) based on their vocabulary score on a developmental test. A stands for the age of the toddler in months. S stands for the toddler's score on the vocabulary test.

$$A = \frac{1}{4}S - 5$$

Estimate the age of a toddler who scores a 100 on the vocabulary test.

This just means that we evaluate the equation when $S = 100$.

$$A = \frac{1}{4}(100) - 5 = 25 - 5 = 20$$

We estimate that the toddler is 20 months old.

Try one on your own.

k) Estimate the age of a toddler who scores a 160 on the vocabulary test.

You may also be asked to interpret the slope of this line. The slope is telling a story. Here, the S is in the place of the x and the A is in the place of the y. Normally, slope is change of y over change of x. Here it would mean:

$$Slope = \frac{\Delta A}{\Delta S} = \frac{change\ in\ age}{change\ in\ score}$$

Our slope was $\frac{1}{4}$, so this would mean "as the score of a child goes up by 4 points, we expect the age of the child to go up by 1 month." Suppose our equation had been each of the following, interpret the slope:

l) $A = \frac{3}{4}S - 5$

m) $A = 2S - 5$ (This is tricky. Remember, the slope could be written as $\frac{2}{1}$.)

We want to look at pairs of lines which have special relationships: parallel lines and perpendicular lines. Parallel lines are like railroad tracks. They are always an equal distance apart and they will never cross. Perpendicular lines make a perfect x, creating 90° angles between the lines.

Using slope-intercept form, graph both of these lines on the same graph:

a)

$$y = \frac{3}{2}x + 5$$

$$y = \frac{3}{2}x + 1$$

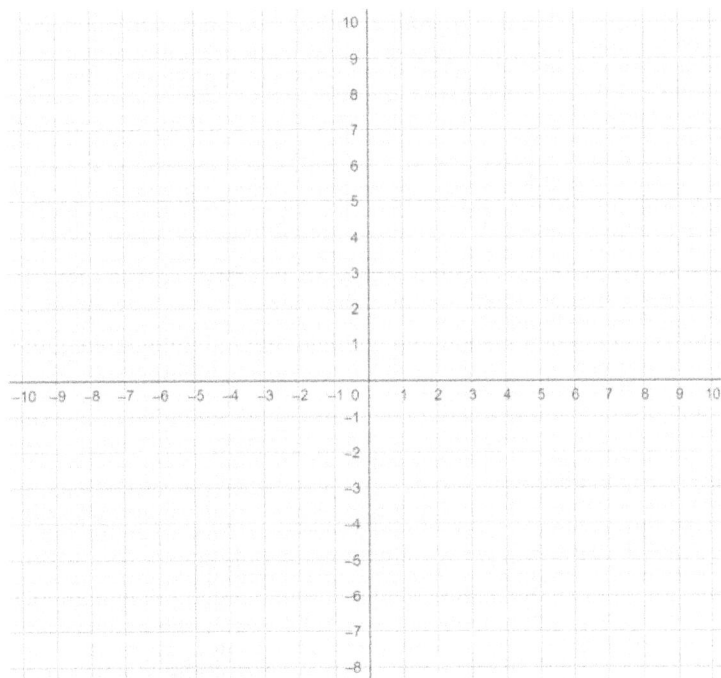

b) Are these parallel lines or perpendicular lines?

c) What in the formula creates this kind of pair?

Graph the following lines:

d)

$$y = -\frac{2}{3}x - 1$$

$$y = \frac{3}{2}x + 2$$

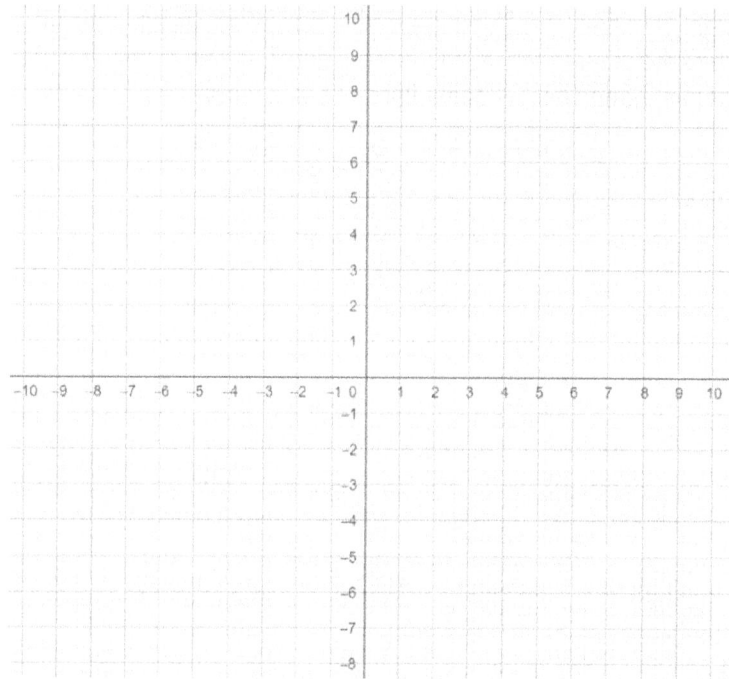

e) Are these lines parallel or perpendicular?

f) Two things about the slope are causing this. What are those two things?

Put the following pairs of equations into slope-intercept form, and indicate if they are parallel, perpendicular, or neither. (Remember, the key is the slope. Keep in mind what you learned above.)

$$7x + 2y = 3$$
$$2x + 7y = 5$$

g) Parallel, Perpendicular, or Neither.

$$3x - 2y = 6$$
$$y = \frac{3}{2}x + 1$$

h) Parallel, Perpendicular, or Neither.

$$4y - 3x = 12$$
$$3y + 4x = 15$$

i) Parallel, Perpendicular, or Neither.

Algebra I

Active Lesson: 4.6a

Find the slope and y-intercept of this line.

a) Slope: $m =$

b) Y-intercept: (0, ___)

c) If we have the slope of the line and the y-intercept, we can simply put those facts together to make the slope-intercept equation for the line. What is the equation of the line above?

d) We don't need to have the y-intercept to find the equation of the line. If we have the slope and any point, we can still find it. Here's how. What is the name of the formula I've written below? (It has been written backward, but it is still something you know.)

$$\frac{y - y_1}{x - x_1} = m$$

I've now rearranged the formula by multiplying the denominator up to the other side.

$$y - y_1 = m(x - x_1)$$

This is called the point-slope form. I know the slope and one point (x_1, y_1). I don't know a second point, so I'm just going to leave them as the variables (x, y). Here is the graph of the same line we already found, but I've marked a new point.

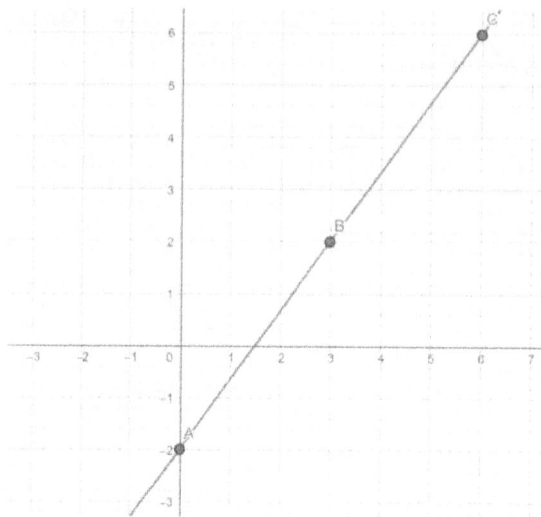

e)   What are the coordinates of the new point C?

f)   What was the slope of this line?

g)   Put the slope and point C into the point-slope equation. Point C is (x_1, y_1).

$$y - y_1 = m(x - x_1)$$

h)   Finally, put the equation into slope-intercept form. (You must first distribute the m and then send y_1 to the other side. Combine any like terms.)

i)   Is your equation the same as the one you originally found above?

j) Find an equation of the line using the point-slope formula. (You must then put it into slope-intercept form.)

$$m = \frac{2}{3}$$

$$(9, 2)$$

Try another.

k) $$m = -\frac{1}{3}$$

$$(6, -4)$$

You could take the idea one step further and find the equation for the line if you only had two points. First, you would need to find the slope. Then, use the slope and either of your points to do a problem like we did above. Once again, let's use the same graph we started with.

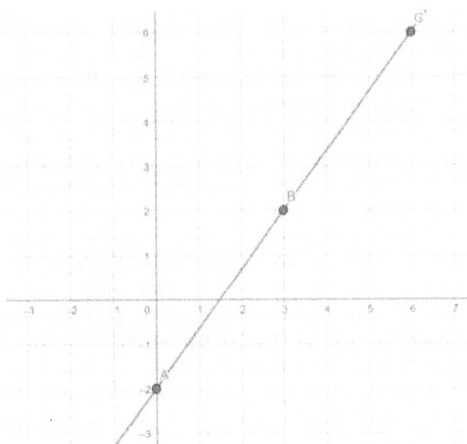

l) What are the coordinates of point C?

m) What are the coordinates of point B?

n) Find the slope between the two points by using the slope formula?

$$m = \frac{y_2 - y_1}{x_2 - x_1}$$

o)   Now, put the slope and point C into the point-slope formula. Then, convert the equation of the line into slope-intercept form.

p)   Next, put the slope and point B into the point-slope formula. Then, convert the equation of the line into slope-intercept form.

If you've done it correctly, you should get the same formula (the one you already knew) using either point B or point C.

q)   Try one more. Find the equation of the line given these two points.

$(3, 2)$

$(5, 6)$

r)   Let's end with a couple tricks. Find the equation of the line given these two points.

$(5, 1)$

$(5, 7)$

(Hint: you could find the slope, but notice that the two x-coordinates are the same. What kind of line would that be?)

s)   And between these two points.

$(4, -1)$

$(15, -1)$

(Hint: Notice that the two y-coordinates are the same. What kind of line would that be?)

Algebra I

Active Lesson: 4.6b

We want to graph a line parallel to the one graphed. Point A is the y-intercept of the line which we want to graph.

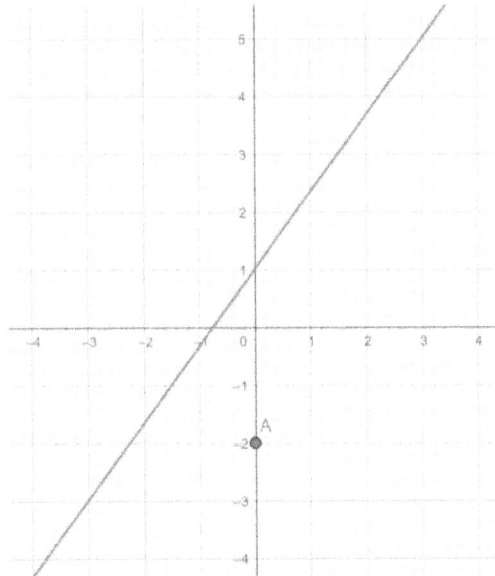

a)   Find the slope of the graphed line.

b)   If our line is parallel to the one already graphed, what do we know about the slope of our line?

c)   What is the slope of the parallel line we want to graph?

d)   Using that slope and point A, graph the parallel line on the graph above.

Let's do the same thing without being able to look at the slope from the graph. We want to graph a line

e) parallel to $y = -\frac{2}{3}x + 3$ with a y-intercept of $(0, -1)$. What is the slope of the line we want to graph?

f) Graph it below.

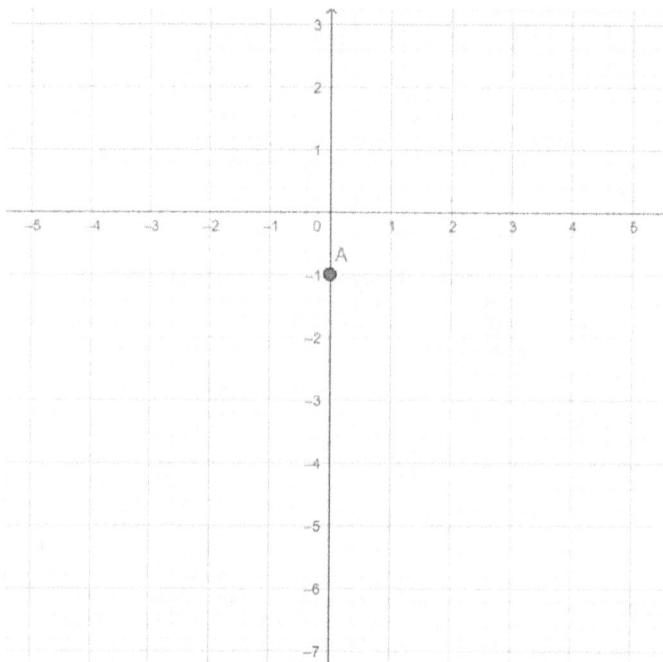

Let's do one more. This time, we aren't going to graph it; we are just going to find the equation for our new line using the point-slope formula.

Find the equation for a line parallel to the line $y = \frac{1}{2}x + 4$ and going through the point $(2, 1)$.

g) What is the slope of our line?

h) Use the point-slope formula with our slope and the point $(2, 1)$. (Rearrange it to put it into slope-intercept form.)

Next, we want to graph a line perpendicular to the one graphed. Point A is the y-intercept of the line we want to graph.

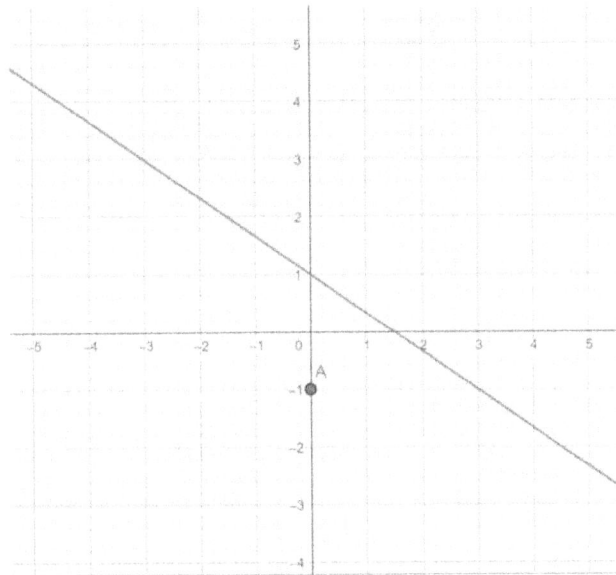

i) Find the slope of the graphed line.

j) If our line is perpendicular to the one already graphed, what do we know about the slope of our line?

k) What is the slope of the perpendicular line we want to graph?

l) Using that slope and point A, graph the perpendicular line on the graph above.

Once again, let's do the same thing without being able to take the slope from the graph.

m) We want to graph a line perpendicular to $y = -3x + 3$ with a y-intercept of $(0, -1)$. What is the slope of the line we want to graph? (Remember -3 is the same as $-\frac{3}{1}$.)

n) Graph it below.

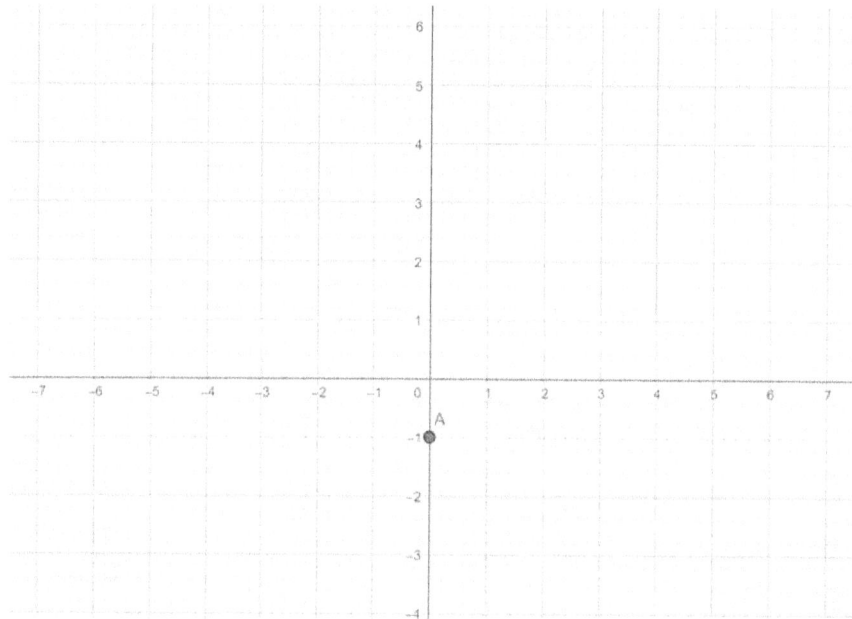

As before, let's find a perpendicular line without graphing. Find the equation for a line perpendicular to the line $y = \frac{1}{2}x + 4$ and going through point $(2, 1)$.

o) What is the slope of our line?

p) Use the point-slope formula with our slope and the point $(2, 1)$. (Rearrange it to put it into slope-intercept form.)

Algebra I

Active Lesson: 4.7

Below is a graph of the inequality $y > \frac{3}{2}x + 2$. I've marked several points. Find the coordinates of the points and put them into the inequality. If the point makes the inequality true, write TRUE. If the point makes the inequality false, write FALSE. Points that make an inequality true are called solutions.

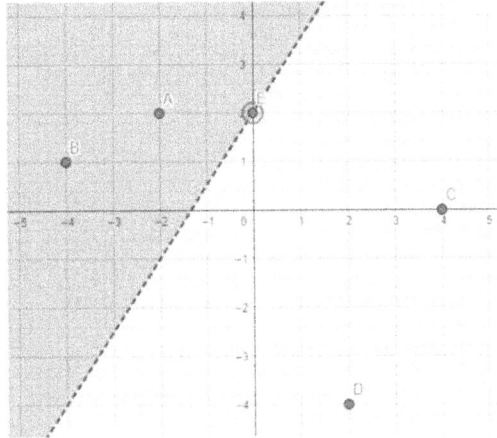

a)

	Coordinates (x, y)	TRUE/FALSE
Point A		
Point B		
Point C		
Point D		
Point E		

b)   What is odd about point E?

c)   Which of these equations would change point E to true: $y < \frac{3}{2}x + 2$ or $y \geq \frac{3}{2}x + 2$

Next, we want to understand what region of the inequality should be shaded.

The key is to look at the y-axis. If your inequality says y is greater than the line, you want the region which includes greater values on the y-axis. This is the graph of $y > \frac{5}{4}x - 1$. We want the region greater than the line.

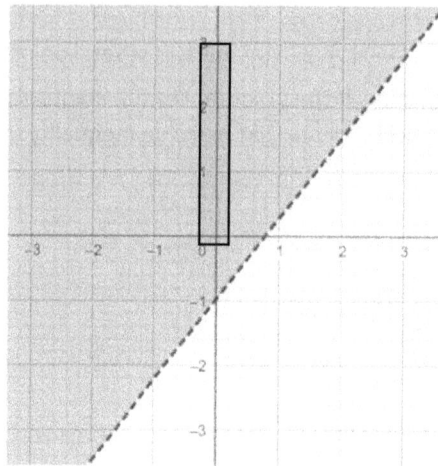

If your inequality says y is less than the line, you want the region which includes lesser values on the y-axis. Here is the graph of $y < -\frac{3}{2}x - 1$.

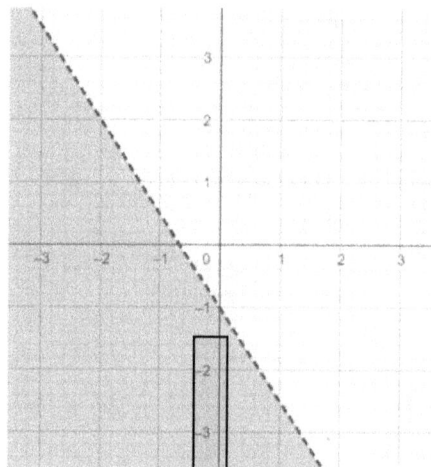

I've created four versions of the same equation by changing the inequality. Match each inequality on the left with its proper graph on the right.

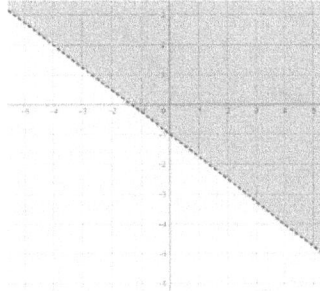

d) $y < -\frac{3}{4}x - 1$

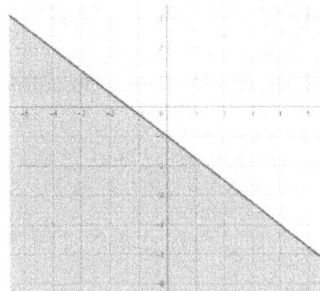

e) $y > -\frac{3}{4}x - 1$

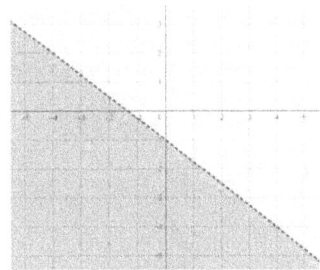

f) $y \leq -\frac{3}{4}x - 1$

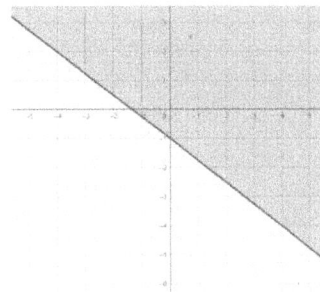

g) $y \geq -\frac{3}{4}x - 1$

h)   Which inequality signs give a dashed line on a graph?

i)   What does the dashed line mean?

Now try graphing.

- At first, ignore the inequality sign and just graph the line like normal. Just be careful to draw the line with either a dashed line or a filled-in line, depending on which is appropriate.
- For greater than inequalities, shade the region above the line. (It should shade the larger values of y on the y-axis.)
- For less than inequalities, shade the region below the line. (It should shade the lesser values of y on the y-axis.)

j)   Graph the following inequality:

$$y < -2x + 3$$

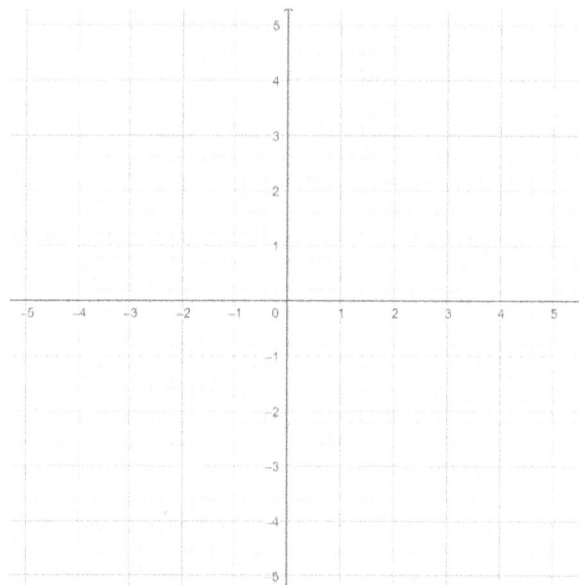

k) Graph the inequality:

$$y \geq \frac{3}{2}x - 2$$

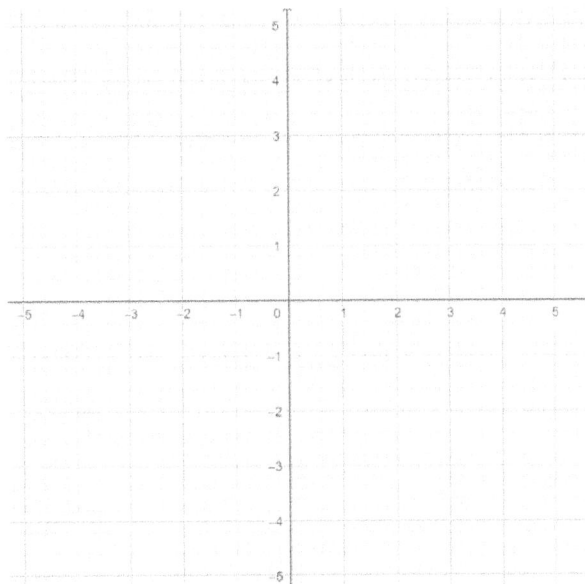

l) Try one more. Be careful, this isn't in slope-intercept form. (Watch out for dividing by a negative. You must flip the inequality.)

$$2x - 3y < 9$$

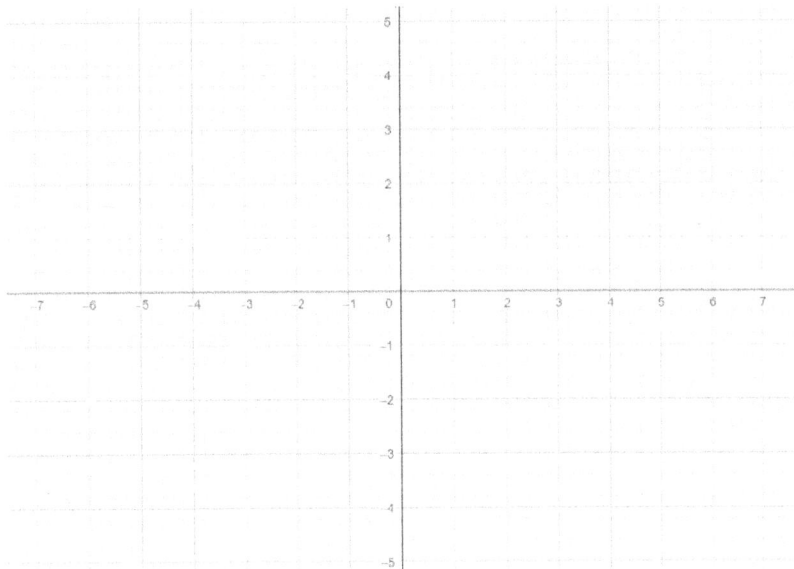

274

In this activity, we want to understand systems of linear equations. A system of linear equations simply means that both lines exist in the same "universe." They exist together, and we are interested in knowing where (if anywhere) they intersect. The point where they intersect is the solution to the system.

One way to find a solution to a system of linear equations is to graph the two lines and see where they intersect. What is the solution to the following systems of linear equations? (Give your solution as a set of ordered pairs.)

$y = -x + 4$

$y = x - 2$

$2x + y = 0$

$3x + y = 1$

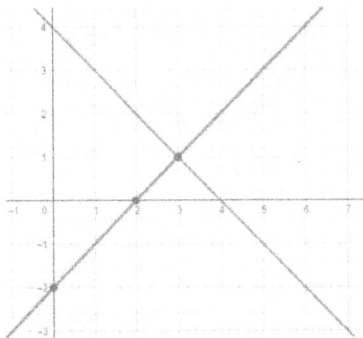

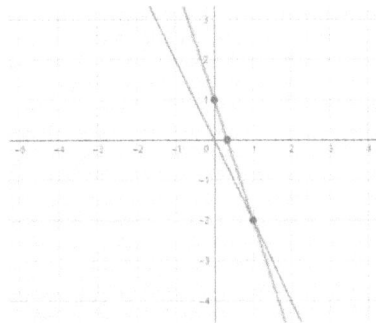

a)   Solution:

b)   Solution:

Graph the following system of linear equations and give the solution.

c)

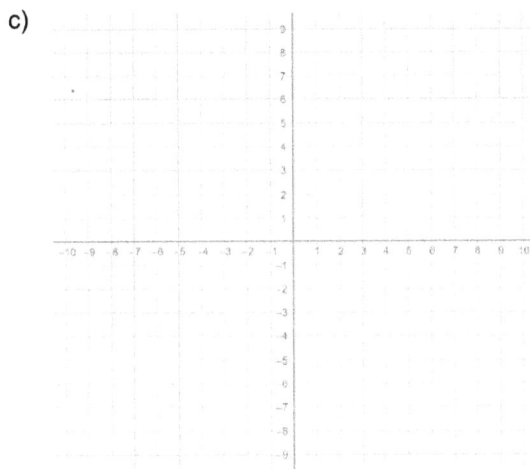

$x - y = -1$

$2x - y = -6$

d)   Graph the following system of linear equations and give the solution.

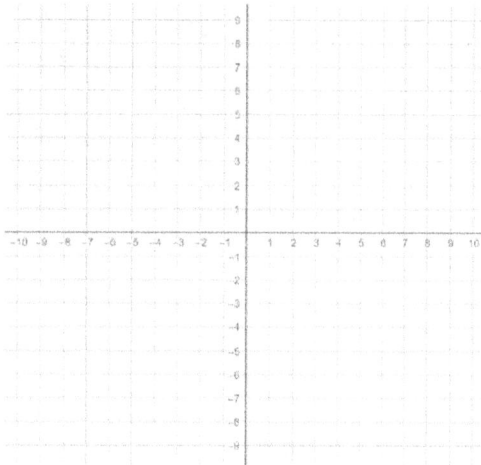

$$y = \frac{1}{2}x + 2$$

$$2x - 4y = -4$$

e)   Something unusual has happened here. What is it?

f)   Why has this happened?

Finally, one of the following two points is a solution to the system of linear equations. Test the two points. The point which gives a true statement in **both** equations is on both lines, and if it is on both lines, it must be where they intersect. And, if it is where they intersect then the point is a solution. (Substitute in for x and y.)

$$y = \frac{1}{3}x + 2$$

$$2x + 8y = 2$$

g) $(6, 4)$ $(-3, 1)$

Algebra I

Active Lesson: 5.2

To understand the next approach that can be used to solve a system of linear equations, solve the following problems by substituting in the known value.

a)   Solve for y if $3x + 2y = 45$ and $x = 5$

b)   Solve for x if $2x - 9y = 82$ and $y = 2$

If we know x or y then solving these equations is easy. The concept for systems of linear equations is the same. We are going to solve for one variable and substitute it into the other equation. Look at this system:

$$2x + y = 0$$

$$3x + y = 1$$

I'm going to solve the top equation for y because it would be easy. (It can be x or y, it doesn't matter.)

$$y = -2x$$

Now I know that y is equal to -2x. It isn't a simple number, but it will be enough to find what I need. Next, I substitute it into the other equation for y.

$$3x + (-2x) = 1$$

And I find $x = 1$.

The final step is to substitute the value of x back into either of the original equations. And, you get the point where they intersect.

$$3(1) + y = 1$$

$y = -2$

$(1, -2)$ is the solution to the system of linear equations. (We graphed this in the last activity and got the same answer.)

Try one. Since x is already solved, substitute it into the other equation. (Always try to choose the easiest variable to solve for. You also may start with either equation. It doesn't have to be the top one.)

$$x = -y - 1$$

$$2x - y = -5$$

You need parenthesis here. I've started it below. You finish it. Give the final answer as an ordered pair.

c) $$2(-y - 1) - y = -5$$

Try another. Solve the system of linear equations.

d) $y = 3x + 2$

$y = 2x$

I've purposely given easy equations. Try one more which is a bit harder. (It would be easy to solve for the y in the top equation or start with the x in the bottom equation. You can do either.)

e) $$2x + y = 7$$

$$x - 2y = 6$$

Do another. Something odd will happen here. Solve the system of linear equations by substitution.

f)

$$x - 5y = 15$$
$$5y = x + 20$$

g) When you substituted your variables fell out. When what remains is false, like should have happened above, there is **no solution**. Why is there no solution? (Graph them if you aren't sure.)

h) One more. Solve the system of linear equations by substitution.

$$2x + 4y = 8$$
$$4x + 8y = 16$$

i) Again, when you substituted your variables fell out. When what remains is true, as should have happened here, there are **infinite solutions**. Why are there infinite solutions? (Graph them if you aren't sure.)

Algebra I

Active Lesson: 5.3

So far, we've learned two methods for solving a system of linear equations: graphing and substitution. In this activity, we will learn a third called elimination. To understand the idea, we need to remember classic elementary school addition.

$$\begin{array}{r} 3 \\ +5 \\ \hline 8 \end{array}$$

Let's add some algebra to the idea. Add the like terms below:

a)
$$\begin{array}{r} 2x \\ +5x \\ \hline \end{array}$$

We can solve linear equations like this too. Look at this system of linear equations, lined up as an addition problem. We can add the like terms. (On the left, add the x's and the y's. Bring down the equal sign. On the right, add the numbers.)

b)
$$\begin{array}{r} 4x + 2y = 14 \\ + x - 2y = 6 \\ \hline \end{array}$$

Now with only one variable, you can solve for x. Solve for x in the space below:

c)

After you solve for x, you can put that value into either of the original equations in order to solve for y. Solve for y:

d)

In this next system, adding the like terms won't get rid of a variable, but there is a simple trick.

$$2x + y = 7$$
$$+\ x - 2y = 6$$

Any equation remains in balance if we multiply each term by the same number. So, if we multiply, the top equation by 2, we now create opposite terms for the y's.

$$2(2x) + 2(y) = 2(7)$$
$$+\ x - 2y = 6$$

And we get:

$$4x + 2y = 14$$
$$+\ x - 2y = 6$$

Adding now removes the y term and we could solve. (This new system is actually the same as the original.) It doesn't have to be the y term that is made to be the opposite. We could multiply through either equation by anything in order to make what we need.

$$2x + y = 7$$
$$+\ x - 2y = 6$$

e) Multiply the bottom equation through by -2 and then add the terms. (You must multiply through every term in the equation, on both sides, to keep the equation in balance.) The x's will now drop out and you can solve for y. In the space below, show that this approach gets the same solution as before.

Try one on your own. You can get rid of either variable, but you must multiply through in order to create an opposite either with the x's or with the y's.

f)
$$x - y = -1$$
$$+\ 2x - y = -6$$

Sometimes you must multiply through both equations to create the opposites. Look at this system.

$$4x - 3y = 9$$
$$+\ 7x + 2y = -6$$

Let's get rid of the y's. However, a single change won't be enough. The trick is to multiply the other equation by the coefficient in front of the y's. Finish the problem below.

g)
$$2(4x) - 2(3y) = 2(9)$$
$$+\ 3(7x) + 3(2y) = 3(-6)$$

Here is another that will require multiplying both equations. (You can eliminate either the x's or y's. However, if you choose the x's, you need to multiply through the top by 5 and the bottom by -3 in order to create the opposites.) Solve the system of linear equations by elimination.

h)
$$3x - 4y = -9$$
$$+\ 5x + 3y = 14$$

Of course, the special cases of **No Solution** or **Infinite Solutions** can occur here too. Solve the following system of linear equations and determine if it should be no solution or infinite solutions.

i)

$$2x - 3y = -4$$
$$+\ 6x - 9y = -12$$

Algebra I

Active Lesson: 5.4a

Earlier, we spent an entire chapter learning how to solve word problems. We would carefully translate from the English language into Math language. The more difficult problems required us to make a dictionary that involved two unknowns; where one of the unknowns could be built from the other.

Now that we know how to solve systems of equations, solving word problems can be much easier. Here, we will begin solving word problems by creating two equations and using the skills of elimination or substitution to solve for two variables.

Let's take a look at the idea:

The sum of the ages of two grown brothers is 119. One brother is 11 years older than the other. What are their ages?

First, we make our dictionary, which is much easier with two variables.

Math	English
x	Age of younger brother
y	Age of older brother

Next, we need to make two equations. One of the equations is typically pretty simple. Here, if we add their ages, we get 119.

$$x + y = 119$$

The second equation is a translation of the sentence, "one brother is 11 years older than the other."

One brother is: (This is the older brother.) $y =$

11 years older than the other: $x + 11$

So:

$$y = x + 11$$

Now, we have a system of linear equations:

$x + y = 119$

$y = x + 11$

a) Use the ideas from the last sections to solve this system of equations. (It is set up well to do substitution.) Give the ages of both brothers by finding both x and y.

Try this problem on your own.

b) Two siblings are 12 years apart. If the sum of their ages is 42, how old are each of the siblings?

Let's look at a second type of problem. The ideas are the same but setting up the equations is a bit different.

Jenni goes to the gym on Monday. She spends 30 minutes on the elliptical and 25 minutes on the bicycle, burning a total of 700 calories. On Thursday, she spends 40 minutes on the elliptical and 15 minutes on the bicycle, burning 750 calories. How many calories does she burn on the elliptical and how many calories does she burn on the bicycle?

Here's our dictionary:

Math	English
x	Rate of Calories Burned on the Elliptical
y	Rate of Calories Burned on the Bike

Here the two equations are built by translation. Since x and y are rates of calories, we can find the total calories by multiplying by the time on each machine.

She spends 30 minutes on the elliptical and 25 minutes on the bicycle, burning a total of 700 calories.

Total calories burnt on the elliptical: 30x

Total calories burnt on the bike: 25y

Next, we need the left side and the right side of the equation to tell the same story. So, both need to tell the total calories burned while exercising.

$$30x + 25y = 700$$

Build the second equation from this sentence:

c) She spends 40 minutes on the elliptical and 15 minutes on the bicycle, burning 750 calories.

d) Use either substitution or elimination to find the values of x and y.

Next, we will work with problems which are based on an idea from geometry. Two complimentary angles are those which add up to 90°. Two supplementary angles are those which add up to 180°. A trick I use to remember is that c comes before s and 90° before 180°.

These problems will require us to know these definitions in order to set up the equations. Here's an example.

The difference of two complimentary angles is 12 degrees. Find the angles.

Here's our dictionary:

Math	English
x	1st Angle
y	2nd Angle

The "easy" equation to set up comes from the definition of complimentary angles. If you add them, they sum up to 90°.

$$x + y = 90$$

The second equation comes from translating, "the difference of two complimentary angles is 12."

$$x - y = 12$$

Find the values of the two angles by any method for solving systems of linear equations.

e) What are the values of the two angles?

Work this problem.

f)   The difference of two supplementary angles is 32 degrees. Find the angles.

Let's look at a final type of problem involving perimeter. To work these problems, we must remember that perimeter is adding up the outside of a shape. We will work with rectangles.

A man wants to put a fence in his back yard. He will not need any fence along the house. It will take 170 feet of fencing. The length of the fence will be 20 feet more than the width. Find the length and the width.

Here's our dictionary:

Math	English
x	Length
y	width

Our first equation is the perimeter. We add up the outside of the rectangle, but there is only one length because of the house.

$$x + 2y = 170$$

Our second equation is the translation of, "the length of the fence will be 20 more than the width."

$$x = y + 20$$

g)   Solve for the length and width. (Notice that this is set up well for substitution.)

Work a perimeter problem on your own.

h)   A man wants to put a fence in his back yard. He will not need any fence along the house. It will take 230 feet of fencing. The length of the fence will be 10 feet less than twice the width. Find the length and the width.

In this section we will look at more word problems. As we did before, our approach will be to solve a system of equations. Here, the problems will be based on a familiar formula:

$$Distance = Rate \cdot Time$$

The ideas will mimic those of the earlier section based on $D = r \cdot t$.

Let's begin by looking at a problem:

Two friends must each drive to a neighboring town. The first friend leaves driving 55 mph. An hour later, the second friend leaves, driving 60 mph. How long until the second friend catches up to the first?

As always, let's begin with the dictionary. It will be the type involving rate multiplied by time.

	Rate	Time	Distance
First Friend	55	x	
Second Friend	60	y	

Remember, the distance column is found by multiplying rate and time. So:

	Rate	Time	Distance
First Friend	55	x	55x
Second Friend	60	y	60y

Next, we make two equations. We know that the two friends drove the same distance, so we can set their distances equal to one another.

$$55x = 60y$$

The second equation is from translating this sentence, "an hour later, the second friend leaves." This is tricky, but it means that the second friend spent 1 hour less in the car. Therefore:

$$y = x - 1$$

Our system of equations is:

$$55x = 60y$$

$$y = x - 1$$

a) Solve the system of equations to find the times. (Notice that the second equation is set up for substitution.)

In the last problem, we didn't know the time for either person. This next type of problem will involve unknown rates. It will typically involve a boat or a plane, and the idea is that the wind (or the river current) can work with the plane, causing it to move faster. Or, the wind can work against the plane, causing it to go more slowly.

A plane can travel 1200 miles in 3 hours with a tailwind (the wind helping) or it can travel 950 miles in 4 hours with a headwind (the wind hurting). Find the rates of the plane and the wind.

So, here is our dictionary:

Math	English
x	Rate of the plane
y	Rate of the wind

But we still need the table with $D = r \cdot t$.

	Rate	Time	Distance
With the Wind	x+y	3	
Against the Wind	x-y	4	

This is unlike any table we've had before. It shows how the wind speeds up the plane and how the wind slows down the plane. As always, the distance column is from multiplying the rate and the time.

	Rate	Time	Distance
With the Wind	x+y	3	3(x+y)
Against the Wind	x-y	4	4(x-y)

To create our two equations, we know that the plane travelled 1200 miles with the wind:

$$3(x + y) = 1200$$

And, the plane travelled 950 miles against the wind:

$$4(x - y) = 950$$

Here is our system:

$$3(x + y) = 1200$$

$$4(x - y) = 950$$

When solving this system, you could use the distributive property. However, since the entire left side is being multiplied by a single number, we can just divide both sides. Here's what I mean.

$$\frac{3(x + y)}{3} = \frac{1200}{3}$$

$$\frac{4(x - y)}{4} = \frac{950}{4}$$

And our two equations become:

$$x + y = 400$$

$$x - y = 237.5$$

Finish solving the system. Notice that it is already set up for the elimination method.

b) What are the speeds of the plane and the wind?

Try this one. The concept is the same.

A boat leaves the dock and can travel 100 miles upstream (working against the river's current) in 5 hours. It takes the boat 4 hours (working with the river's current) to come back to the dock.

Our dictionary looks like this:

Math	English
x	Rate of the boat
y	Rate of the river current

Here is the beginning step in the distance table. Complete the distance column.

	Rate	Time	Distance
With the Current	x+y	4	
Against the Current	x-y	5	

c) Finish the problem by creating a system of equations. The key is that the distance the boat travelled upstream and the distance the boat travelled downstream were both 100 miles.

Algebra I

Active Lesson: 5.5a

Just as we had previously worked with distance problems, the word problems in this section have been covered too. This time, we will investigate them using a system of linear equations.

A school play sold 1250 tickets. They made a total of $9,750. How many adult tickets and how many student tickets were sold if adult tickets cost $12 and student tickets cost $6?

As always, we start with our dictionary.

Math	English
x	Adult Tickets Sold
y	Child Tickets Sold

Now, we need to find two equations. The easy equation is the total number of tickets sold.

$$x + y = 1250$$

Next, we need an equation about the money made. Remember, the right side and the left side must tell the same story:

$$12x + 6y = 9750$$

If we check to make sure both sides are correct, we must have the total money made from the tickets on each side. So, this is correct, and we have this system of equations:

$x + y = 1250$

$12x + 6y = 9750$

a) Solve the system of equations and give both the number of adult tickets sold and the number of children's tickets sold.

Try this problem on your own.

b) The community theater sold 1550 tickets to their latest show. They made a total of $27,200. If adult tickets cost $25 and children's tickets cost $10, how many of each type of ticket did they sell?

Next, we will revisit problems involving coins.

You have a pocketful of nickels and dimes. The total value of your coins is $7.00. The number of dimes is 5 more than twice the number of nickels. Here's our dictionary.

Math	English
x	# of nickels
y	# of dimes

We don't have the total number of coins, so we need to make an equation about the total amount of money. If you remember this from earlier, we need to multiply the number of each coin by their value.

$$.05x + .10y = 7.00$$

Check to make sure both sides of the equation tell the same story. It does. We have the total money on each side. Our next equation is a translation: the number of dimes is 5 more than twice the number of nickels.

$$y = 5 + 2x$$

And our system becomes:

$.05x + .10y = 7.00$

$y = 5 + 2x$

I would clear out the decimals in the first equation by multiplying each group by 100. It won't change how the problem works. If we do that, our system becomes:

$5x + 10y = 700$

$y = 5 + 2x$

c) Finish solving for the number of each coin. (You are set up nicely for substitution.)

Try this problem.

d) A jar filled with quarters and dimes has a total value of $12.55. If the number of quarters is 5 less than twice the number of dimes, find the number of each type of coin.

Next, we revisit problems involving mixtures.

You need 15 pounds of a mixture of cereal and raisins. You need the mix to cost $5.40 per pound. If the cereal costs $5 per pound and the raisins cost $7 per pound, how many pounds of each do you need?

Our dictionary:

Math	English
x	Lbs of cereal
y	Lbs of raisins

Our easy equation would be the total number of pounds.

$$x + y = 15$$

This tells the same story on each side of the equation. Our second equation needs to be about the cost.

$$5x + 7y = 5.40(15)$$

The left side is the total cost of the mix so the right side must be the same. For that reason, I multiplied the 5.4 by the 15 pounds. This gives the total price of our final mix.

The system of equations then becomes:

$$x + y = 15$$
$$5x + 7y = 81$$

e) Solve the system to find the pounds needed of both the cereal and the chocolate chips.

Try another.

f) Eight pounds are needed of a mixture of nuts. The mixture must cost $8.50 per pound. If the mixture is made up of cashews, costing $7 per pound, and almonds costing $10 per pound, find how many pounds of each nut is needed.

g) Our final type of problem is mixtures of liquids. To understand such problems, first answer the following questions:

   a) A 100 ml liquid is 12% acid. How many ml of acid are there?

   b) A 55 ml liquid is 35% acid. How many ml of acid are there?

Now, let's look at a problem.

During a Chemistry lab, an 80 ml mixture of 20% acid must be made. How much of a 10% acid solution and a 35% acid solution should be mixed to create the 20% mixture which is required?

The dictionary looks like this:

Math	English
x	mls of 10%
y	mls of 35%

We are going to pour two liquids together to get an 80ml mixture.

$$x + y = 80$$

Now, we need the amount of acid.

$.1x$ is the amount of acid in the first bottle.

$.35y$ is the amount of acid in the second bottle.

$.2(80)$ is the total amount of acid in the new mixture.

Our second formula must be the total amount of acid on both the left and right.

$$.1x + .35y = .2(80)$$

So, our system of equations becomes: (I've multiplied the right side of the second equation.)

$$x + y = 80$$

$$.1x + .35y = 16$$

h) Solve this system of equation for the mls of each liquid that we would need to pour together to make the new mixture.

Work this problem.

i) A 70 ml mixture of 40% sulfuric acid is needed. How many mls of 35% mixture and a 55% mixture should be combined to make the new 40% mix?

Algebra I

Active Lesson: 5.5b

Next, we are going to work with interest word problems. The idea is based on the simple interest formula:

$$I = Prt$$

But we will be assuming time is equal to one, which just leave $P \cdot r$. P stands for the principal and r stands for the rate. So, as we've done with the mixture problems, we will multiply by a percentage changed to a decimal.

You have $50000 to invest. There is a CD which is yielding 3.5% interest and a mutual fund earning 7.5% interest. If your goal is to earn a combined 5% interest, how much should you invest in each fund?

This is exactly like the mixture problems from the last section. Here is our dictionary:

Math	English
x	Amount in 3.5%
y	Amount in 7.5%

As before, we need two equations for the two variables. The first is easy. We just add up the money.

$$x + y = 50000$$

Next, similarly to the mixture problems, we multiply each amount by the interest rate:

$$.035x + .075y = 50000(.05)$$

Notice how both sides of the equation match. The left side is money earned from each fund, so the right side needs to be the same.

Our system of equations then becomes:

$x + y = 50000$

$35x + 75y = 2500000$ (I've cleared the decimals by multiplying through this equation.)

a)   Finish solving the system of equations to find the amount to invest in the 3.5% fund and the amount to invest in the 7.5% fund.

Work this problem on your own.

b) A finance student has \$30,000 to invest. A CD has a 4% interest rate and a stock portfolio has an 8% interest rate. If the student wants to earn an overall rate of 6.5% how much should they invest in each fund.

Next, instead of an investment making money, we will look at loans. The concept with loans is the same. The person taking out the loan is losing money, but from the bank's perspective, they are earning money. Here's a problem.

A student has \$20,500 in loans. One loan charges 8.5% and the other charges 12.5%. If the total amount of interest the student paid was \$2,262.50, what was the principal in each loan?

The principal is simply the amount of money for each. Here's the dictionary:

Math	English
x	Amount in 8.5% Loan
y	Amount in 12.5% Loan

First, the easier equation is the total amount of money:

$$x + y = 20500$$

The second equation is based on the interest. Here, they have given us the total interest paid. So, look how the right and left sides of the equation tell the same story.

$$.085x + .125y = 2262.50$$

c) Finish solving the system. I'd recommend multiplying the second equation through by 1000.

Finally, try one on your own.

d) A student has two loans for $32,500. One loan has an interest rate of 8%. The second loan has an interest rate of 13%. If the total amount of interest the student paid was $3,475, find the amount of principal in each loan.

Earlier, we learned to graph linear inequalities. Now, we are going to graph systems of linear inequalities. Below is a system of linear inequalities and their graph. (Remember, this means they exist in the same universe.)

$$y \geq 3x + 1$$

$$y < -\frac{1}{2}x - 2$$

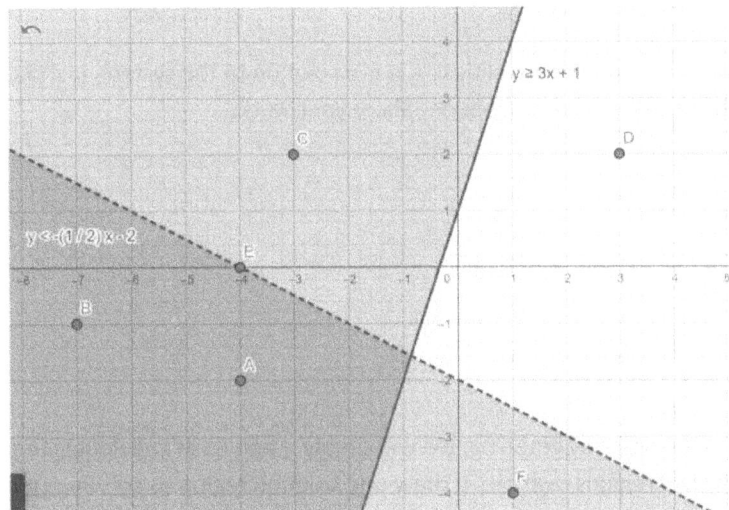

Each inequality creates its own solution region. But, in order for a point to be a solution to the system of equations, it must make both equations true. Two points above are a solution to the system of linear

a) inequalities. What points are they?

b) Point E is not a solution. Why?

c) What are the coordinates of point A?

Test point A in the system of linear inequalities. If it is a solution to the system, it should make both
d) equations say something which is true. Does it? Show your work.

e) What are the coordinates of point C?

Test point C in the system of linear inequalities. If it is a solution to the system, it should make both
f) equations say something which is true. Does it? Show your work.

To solve a system of inequalities by graphing, we will simply graph each individual inequality and see
where their two individual regions overlap. (I draw one solution region as soft lines going in one
direction and the other as soft lines going in a different direction. This causes the overlap region to be a
grid. Or, if you have some, you can use colored pencils or different colored pens.)

g) Solve the system of linear equalities by graphing:

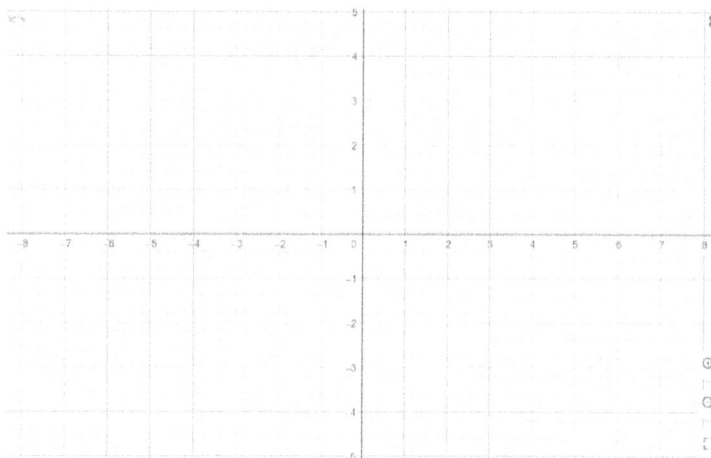

$$y < \frac{2}{5}x + 2$$

$$y \geq -\frac{1}{3}x - 1$$

Sometimes the overlap regions are a bit harder. In this next problem, both inequalities are less than.

h) Remember, you must choose the regions which makes them **both** true.

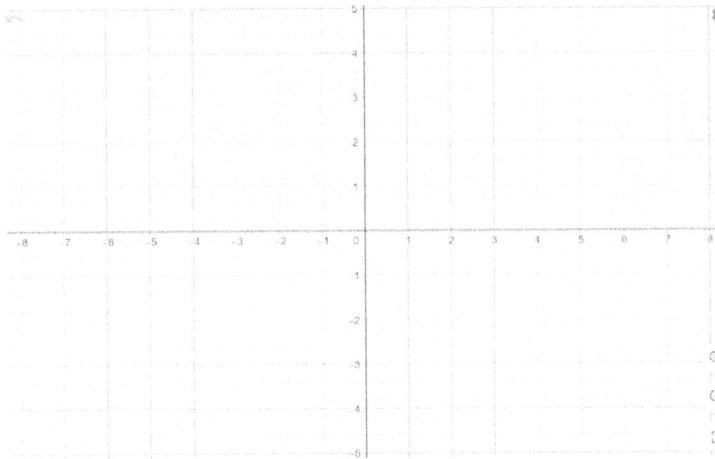

$$y < \frac{2}{5}x + 4$$

$$y < \frac{1}{3}x - 1$$

i) Finally, try a variation.

$$y > \frac{3}{4}x - 2$$

$$x < 4$$

(The solution region to the second problem is the region left of the horizontal line $x = 4$. This is where the x-axis is less than the line.)

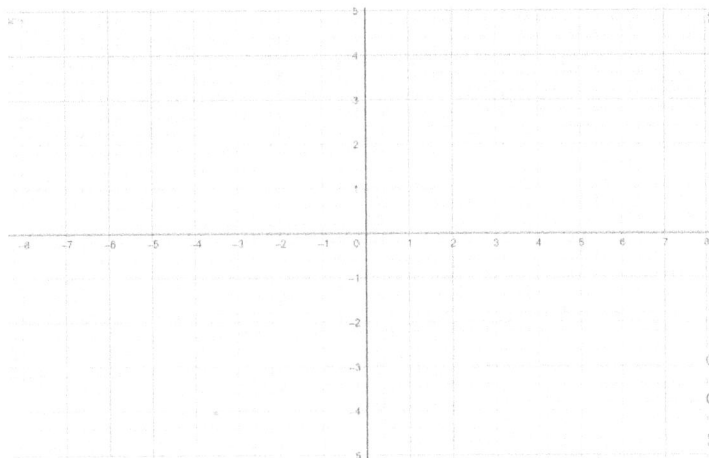

Algebra I

Active Lesson: 5.6b

In our last activity, we graphed systems of linear inequalities. Here, we want to look at applications of those systems.

The team dietician has set a calorie goal for you. You need to eat at least 850 more calories before practice. A cup of milk has 150 calories and a granola bar has 220. A cup of milk costs $1.50 and a granola bar cost $2. You can't spend more than $15 on food.

As with any word problem, let's start with the dictionary.

Math	English
x	# of cups of milk
y	# of granola bars

There are two inequalities here. One inequality is about calories. The other inequality is about the cost. Let's start with the calories.

$$150x + 220y \geq 850$$

Both the left and right sides are calories consumed, so we have a match. The second equation is about the cost.

$$1.50x + 2y \leq 15$$

Here is a graph of the system of linear inequalities:

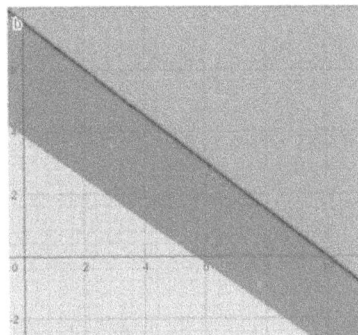

a) Remember, x is cups of milk and y is granola bars. Circle any of the following answers which would meet both the calorie goal and the financial goal: (An answer must be in the overlap region to be correct.)

   a) (2 cups of milk, 2 granola bars)
   b) (8 cups of milk, 1 granola bar)
   c) (6 cups of milk, 4 granola bars)
   d) (4 cups of milk, 2 granola bars)

Let's look at another.

For a bake sale, you would like to sell at least 10 cakes. Chocolate cakes cost $6 to make and Vanilla cakes cost $5.50. You can't spend more than $50 on the cakes.

Create the system of linear equations that represents this system. Here is your dictionary:

Math	English
x	# of chocolate cakes
y	# of vanilla cakes

b)   For the first equation, add the number of each type of cake. You must be greater than or equal to 10.

c)   The second equation is about the money. Both sides of your equation must represent the total cost.

Here is a graph of the correct system of linear inequalities:

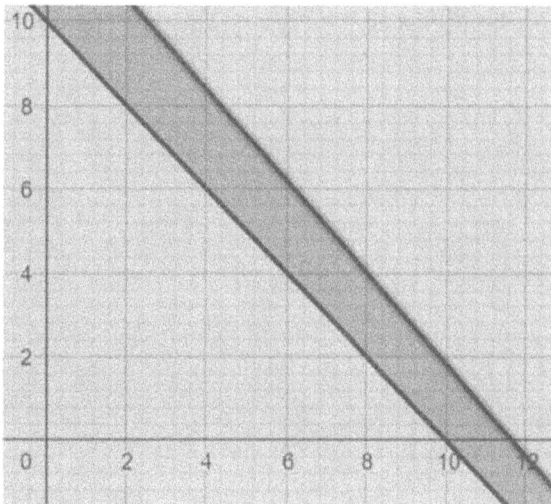

d)   Remember, x is chocolate cakes and y is vanilla cakes. Circle any of the following answers which would meet both the quantity and the financial expectations.

      a) (2 chocolate, 7 vanilla)
      b) (4 chocolate, 10 vanilla)
      c) (6 chocolate, 5 vanilla)
      d) (8 chocolate, 3 vanilla)

Polynomial is a fancy name for a special type of algebraic expression. A polynomial is made up of different groups (called terms) each separated by addition or subtraction.

$$4x^3 + 2x^2 + 4x - 3$$

This polynomial has four terms.

There are special names for some smaller polynomials. One term is called a monomial: $3xy^2$. Two terms is called a binomial: $12y^4 - 2y^2$. Three terms is called a trinomial: $x^3 - 5x^2 + 8x$. And if it exceeds three terms, we simply called it a polynomial. Name each of the following polynomials based upon its number of terms:

a) $15x^7 - 30x$

b) $12x^3y - 20x^2y^2 + 6xy^3 - 50$

c) $3x^3y^2z$

d) $x^2 - 6x + 15$

A polynomial can also be classified by its highest exponent, called the degree.

$$4x^3 + 2x^2 + 4x - 3$$

The highest exponent here is three, so this is a polynomial of degree three. In the real world, it isn't common to have multiple variables in a single term. However, when we see it, the degree is the largest sum of exponents in a term:

$$2x^3y^2 + 7x^2y + 4xy - 53$$

This polynomial has a degree of 5. The first term has an exponent of 3 on the x and 2 on the y.

$(3 + 2 = 5)$

Give the degree of the following polynomials:

e) $15x^7 - 30x$

f) $12x^3y - 20x^2y^2 + 6xy^3 - 50$

g) $3x^3y^2z$

h) $x^2 - 6x + 15$

Adding and subtracting polynomials is nothing more than combining like terms. Remember, the variables are like words. You can combine terms if the words are identical.

$$(2x^2 - 3x + 5) + (3x^2 + 10x - 15)$$

The x^2 terms may combine. The x terms may combine. And the numbers can combine.

$$(2x^2 + 3x^2) + (-3x + 10x) + (5 - 15)$$

$$5x^2 + 7x - 10$$

Add the following polynomials:

i) $(8x^2 + 13x - 6) + (4x^2 - 10x + 23)$

j) $(4y^3 - y - 7) + (13y^2 + 7y + 8)$ (Careful, not all the terms combine here.)

k) $(2x^2y + 3xy^2 + 5xy + 6) + (3x^2y + 10xy - 5)$ (Terms only combine if the variables are identical.)

Subtracting polynomials has only one difference. The difference depends on the distributive property. Multiply the following:

l)
$$-1(2x^2 + 5x - 6)$$

A multiplication sign in front of a polynomial works just like -1. Multiply the following:

m)
$$-(2x^2 + 5x - 6)$$

When subtracting polynomials, the subtraction sign must be distributed through the second polynomial. Then, you can combine like terms.

$$(2x^2 - 3x + 5) - (3x^2 + 10x - 15)$$
$$2x^2 - 3x + 5 - 3x^2 - 10x + 15$$
$$(2x^2 - 3x^2) + (-3x - 10x) + (5 + 15)$$
$$-x^2 - 13x + 20$$

Subtract the following polynomials.

n) $(8x^2 + 13x - 6) - (4x^2 - 10x + 23)$

o) $(4y^3 - y - 7) - (13y^2 + 7y + 8)$

p) $(2x^2y + 3xy^2 + 5xy + 6) - (3x^2y + 10xy - 5)$

Algebra I

Active Lesson: 6.1b

In this activity, we want to evaluate a polynomial. Evaluating is easy. Just substitute a value in for a variable. But when you sub-in you put the value in a parenthesis.

Evaluate when $x = -2$.

$4x$

$$4(-2) = -8$$

The parenthesis are important so that you don't overlook any multiplication. Work the following:

Evaluate when $y = -3$

a) $-5y^2$ (Remember to follow the order of operations. Square the -3 first; then multiply by -5)

To evaluate a polynomial, we just substitute in place of all the variables.

Evaluate when $a = 5$

$2a^2 - 2a + 3$

I will make the substitution and then you simplify. (Remember the order of operations!)

b) $2(5)^2 - 2(5) + 3$

Use the following polynomial to evaluate each of the following:

$$3x^2 - 2x + 10$$

c)   Evaluate when $x = 4$

d)   Evaluate when $x = -3$

e)   Evaluate when $x = 0$

The idea can easily be extended to evaluate multiple variables at the same time.

$$5x^2y - 2xy^2$$

f)   Evaluate when $x = 1$ and $y = 2$

g)   Evaluate when $x = -2$ and $y = -1$

h)   Evaluate when $x = 0$ and $y = 10$

Algebra 1

Active Lesson: 6.2a

Look at the following:

$$2 \cdot 2 \cdot 2 \cdot 2$$

a) Which of the choices below is the same:

    a) 4^2
    b) 2^4
    c) 2^5

Things work the same with variables.

$$x \cdot x \cdot x$$

b) Which of the choices is correct:

    a) x^3
    b) x^4
    c) x^2

Exponents are repeated multiplication. Simplify these:

c) $y \cdot y$

d) $v \cdot v \cdot v \cdot v \cdot v \cdot v$

Next, look at this:

$$x^2 \cdot x^3 = x \cdot x \cdot x \cdot x \cdot x$$

e) Which is the same:

    a) x^4
    b) x^5
    c) x^6

Try the following:

f) $x^5 \cdot x^4$

g) $y^4 \cdot y^3 \cdot y^2$

When an exponent isn't shown, it is like a 1. Try these:

h) $x^3 \cdot x = x^3 \cdot x^1$

i) $y^6 \cdot y^3 \cdot y$

When you have exponents which all have the same base, what mathematical operation are you doing?

j)   (Circle the correct answer.)

   a) Add
   b) Subtract
   c) Multiply
   d) Divide

Now let's see what would happen if you have an exponent to an exponent: $(x^3)^2$. To understand what needs to be done, simply expand it. Since it is squared, you have two x^3's.

$$(x^3)^2 = x^3 \cdot x^3$$

But we already know how to handle that. Give the answer:

k) $(x^3)^2 = x^3 \cdot x^3 =$

Simplify the following:

l) $(y^4)^3 = y^4 \cdot y^4 \cdot y^4 =$

m) $(x^5)^2 =$

n) $(a^{10})^4 =$

There is nothing wrong with expanding them, but after doing them over and over, a pattern begins to emerge. Look at the following and then compare them to your answer from above:

$(y^{4})^{3}$ $(x^{5})^{2}$ $(a^{10})^{4}$

o) When you have an exponent to an exponent, what mathematical operation can you use as a shortcut? (Circle the correct answer.)

    a) Add
    b) Subtract
    c) Multiply
    d) Divide

You can extend the idea if there are multiple variables inside a parenthesis. Just give the exponent to each. Simplify the following:

p) $(x^3 y^2)^2 = (x^3)^2 (y^2)^2 =$

Try these:

q) $(m^5 n^4)^3 =$

r) $(x^3y^6z)^2 =$

And if there is a coefficient (a number) inside, share the exponent with the number.

$(2x^3)^2 = (2^2)(x^3)^2 = 4x^6$

Work these:

s) $(3x^2)^3 =$

t) $(4m^2n^3)^2 =$

Simplify the following using exponent rules:

a) $x \cdot x \cdot x \cdot x =$

b) $y^2 \cdot y^3 =$

c) $a^5 \cdot a^7 \cdot a =$

If there are numbers in front of the variable (coefficients), we first multiply the numbers and then combine the variables.

$$2x^2 \cdot 5x^3 = 10x^5$$

Try these:

d) $6x^3 \cdot 4x^5$

e) $10y^5 \cdot 5y^3$

f) $2x \cdot 7x^8$

g) $4y^2 \cdot 6y^3 \cdot y$

h) $-3x \cdot 8x^2$

i) $-5x \cdot -3x$

We can extend the idea by multiplying a monomial by a polynomial. First, a quick refresher on the distributive property. Simplify the following:

j) $3(x - 5)$

k) $2(5x^2 + 6x - 2)$

l) $6(4xy + 6)$

When we multiply a monomial (one term) by a polynomial (multiple terms), we use the distributive property.

$$4x(3x + 2) = (4x)(3x) + (4x)(2) = 12x^2 + 8x$$

Simplify these:

m) $5x(10x - 3)$

n) $2y^2(4y + 7)$

o) $3a(8a^2 - 2a + 5)$

p) $4xy(9xy - 2)$

q) $7x^2y(2x^2y + x - 1)$

r) $-2x^2y(x^2 + 5xy + y^2)$

Multiply the following:

a)
$$12x(x-5)$$

b)
$$4x^2(x^2-2x+3)$$

To multiply a binomial by a binomial, the idea behind these problems is the distributive property. Simplify the following:

c)
$$x(x+3)$$

d)
$$2(x+3)$$

So, to multiply a binomial by a binomial, we are just going to do the distributive property twice.

$$(x+2)(x+3) = x(x+3) + 2(x+3)$$

Finish this by multiplying and combining like term:

e)
$$x(x+3) + 2(x+3) =$$

Try these. Do the distributive property twice:

f) $(2x+5)(3x+2)$

g) $(5y - 2)(10y + 4)$

h) $(7a + 3)(a^2 - 1)$

i) $(2ab - 1)(3ab + 1)$

j) $(x - 3)(x + 3)$

And, of course, the idea can be extended further.

$$(x + 5)(2x^2 + 4x - 2)$$

You can distribute three times. Here it is in parts. Simplify each part.

k) $x(2x^2 + 4x - 2)$

l) $5(2x^2 + 4x - 2)$

Now, combine the like terms from each of your two answers.

m)

Try these:

n) $(x + 3)(x^2 + 5x + 1)$ o) $(x - 1)(x^2 + 3x - 1)$

Algebra I

Active Lesson: 6.4a

In this lesson, we will look at a special outcome which happens when we multiply some binomials. It isn't necessary to memorize it. You begin to recognize it simply from repetition. Multiply the following:

a) $(x + 3)(x + 3) =$

b) $(y - 2)(y - 2) =$

c) $(2x + 5)(2x + 5) =$

d) What is true about the two binomials we are multiplying?

With these binomials, you always create a trinomial. And, here's the pattern.

1) Square of the first term.
2) Square of the last term.
3) Two of the first term times the last term.

This is true, but you don't need to memorize it. Just multiply it out. You will always get the same result. Try a couple more:

e) $(2x - 3)(2x - 3)$

f) $(x^2 - 1)(x^2 - 1)$

g) $(x - 6)^2$

h) $(2x - 3y)(2x - 3y)$

i) $(2m^2 + 8)^2$

Algebra I

Active Lesson: 6.4b

In the last activity, we saw a special product which occurs whenever we multiply two of the same binomials. In this activity, we are going to find another special product. Again, it doesn't need to be memorized, although after doing enough of them, you will begin to know the pattern by heart.

Multiply the following:

a) $(x + 3)(x - 3) =$

b) $(y - 2)(y + 2) =$

c) $(2x + 5)(2x - 5) =$

d) What is true about the two binomials we are multiplying?

e) After you have multiplied them, what is always true of the sign in the middle?

In this special case, you always create a trinomial. And, here's the pattern.

1) Square of the first term.
2) Square of the last term.
3) Two opposite middle terms which cancel out.
4) In the final answer, the sign between the first and last term is always negative.

Again, this is true, but you don't need to memorize it. Just multiply it out. You will always get the same result. Try a couple more:

f) $(2x - 3)(2x + 3)$

g) $(x^2 + 1)(x^2 - 1)$

h) $(2x - 3y)(2x + 3y)$

i) $(2m^2 + 8)(2m^2 - 8)$

What we are multiplying here are called conjugates. The two terms are the same but they have opposite signs in-between. Conjugates are useful because when conjugates are multiplied the middle term will always drop out. Give the conjugate for each of the following:

j) $(x + 9)$

k) $(2y - 7)$

l) $(a^2 + b^2)$

Algebra I

Active Lesson: 6.5a

There are lots of "rules" about exponents, particularly related to exponents and fractions. But they have an easy-to-understand explanation. The shortcuts are helpful, but before you blindly memorize, just notice where they come from. Here's the first idea. Just expand.

$$\frac{x^5}{x^3} = \frac{x \cdot x \cdot x \cdot x \cdot x}{x \cdot x \cdot x}$$

If you have a matching factor on top and on bottom, you can cancel:

$$\frac{x^5}{x^3} = \frac{x \cdot \cancel{x} \cdot \cancel{x} \cdot x \cdot x}{\cancel{x} \cdot \cancel{x} \cdot \cancel{x}}$$

There are now two x's on top, so the answer is x^2.

Expand these, and give the final answer:

a) $\dfrac{x^7}{x^4} = \dfrac{x \cdot x \cdot x \cdot x \cdot x \cdot x \cdot x}{x \cdot x \cdot x \cdot x} =$

b) $\dfrac{y^3}{y^2} =$

c) $\dfrac{a^8}{a^3} =$

Instead of expanding and cancelling, someone began to notice a pattern.

$$\frac{x^5}{x^3} = x^2$$

d)   What mathematical operation can we do with the exponents: (Circle the correct answer.)

    a)   Addition
    b)   Subtraction
    c)   Multiplication
    d)   Division

It is also possible to have more variables left on the bottom (denominator).

$$\frac{x^3}{x^5} = \frac{x \cdot x \cdot x}{x \cdot x \cdot x \cdot x \cdot x} = \frac{1}{x^2}$$

Try these:

e) $\dfrac{x^2}{x^6} =$

f) $\dfrac{y^7}{y^{10}} =$

If we could subtract exponents before, we can do it again, and an odd thing happens.

$$\frac{x^3}{x^5} = x^{-2}$$

And this is okay. It simply must mean that:

$$\frac{1}{x^2} = x^{-2}$$

g)   Another odd thing can occur. What is the answer to the following:

$$\frac{2}{2} =$$

In the same way, suppose:

$$\frac{x^3}{x^3} = \frac{x \cdot x \cdot x}{x \cdot x \cdot x}$$

h)   Cancel the x's. If everything cancels on the top and bottom, what must the answer be?

But do the subtraction trick with the exponents and you get $x^{3-3} = x^0$.

We can't do the same problem two ways and get two different answers, and so, the following must be true: $x^0 = 1$. Answer the following. (Don't over think them. It is as easy as it appears.)

i) $x^0 =$

j) $2^0 =$

k) $(xy)^0 =$

l) $\left(\frac{x}{y}\right)^0 =$

m) $(3x^3y^5)^0 =$

We have one last idea. Finish this problem:

$$\left(\frac{3}{4}\right)^2 = \frac{3}{4} \cdot \frac{3}{4} =$$

Exponents are repeated multiplication. Now finish this problem:

$$\frac{3^2}{4^2} =$$

You get the same thing. Here's the point. If you have a fraction with an exponent, you can give the exponent to the top (numerator) and to the bottom (denominator).

Try these:

n) $\left(\frac{2}{5}\right)^2 =$

o) $\left(\frac{-2}{y}\right)^2 =$

p) $\left(\frac{4}{x}\right)^3 =$

q) $\left(\frac{a}{b}\right)^6 =$

In this activity, we will look at more complex problems involving the rules for exponents. Here's the basic strategy. Clean up the top. Clean up the bottom. Finally, divide.

$$\left(\frac{x^4}{x^3}\right)^2$$

Give the square to the top and bottom:

$$\frac{(x^4)^2}{(x^3)^2}$$

a) Now clean up the top. Then, clean up the bottom. Finally divide. Show your work below.

You should have gotten: x^2.

Try another. Show your work.

b) $\left(\frac{y^2}{y^3}\right)^3$

You could have written your answer as $\frac{1}{y^3}$ or y^{-3}. Try these:

c) $\left(\frac{a^6}{a^4}\right)^3 =$

d) $\dfrac{x^6}{(x^2)^3} =$

e) $\left(\dfrac{x^4}{y^3}\right)^2 =$

In this last problem, you could clean up the top and bottom, but there wasn't anything that could divide. That's okay. Just simplify the best you can. Try this problem:

f) $\left(\dfrac{a^5}{b^3}\right)^4 =$

On this next one, be sure that you remember to square the 2 on top as well.

g) $\left(\dfrac{2x^2}{y^3}\right)^2 =$

Now, you must also square the 2 on top and the three on the bottom.

h) $\left(\dfrac{2x^3}{3y^2}\right)^2 =$

Things can get more complicated, but the strategy stays the same.

i) $\dfrac{(x^2)^3(x^3)^2}{(x^3)^5} =$

You should have gotten $\dfrac{1}{x^3}$ or x^{-3}. Try some more problems:

j) $\dfrac{(y^4)^5}{(y^3)^2(y^4)^2} =$

k) $\dfrac{(2x^2)^2(3x^3)^2}{(x^3)^2} =$

l) $\dfrac{(2y^2)^2}{(3y^2)^2(y^3)^2} =$

Algebra I

Active Lesson: 6.5c

Simplify the following:

a) $\dfrac{2}{6} =$

b) $\dfrac{x^2}{x^3} =$

c) $\dfrac{y^5}{y^3} =$

Individually, those were each rather easy. Putting them all together doesn't change anything. You simplify each individual part as if it were three separate problems. Simplify:

d) $\dfrac{2x^2y^5}{6x^3y^3} =$

Try these.

e) $\dfrac{6x^5y^3}{8x^2y^5} =$

f) $\dfrac{-24a^{15}b^{13}}{8a^{12}b^{9}} =$

In these final problems, clean up the top, clean up the bottom, and then divide.

g) $\dfrac{(3x^{2}y^{3})(15x^{5}y^{3})}{(5x^{2}y^{2})} =$

h) $\dfrac{(-6a^{5}b)(-3a^{3}b^{4})}{(2a^{4}b^{2})} =$

Algebra I

Active Lesson: 6.6a

a) Put the following line in slope-intercept form.

$$3x + 2y = 8$$

During the process of putting the equation in slope-intercept, you had to do today's concept—dividing a polynomial by a monomial. You already know all the ideas involved.

Divide the following:

b) $\dfrac{8x^2}{2x}$

c) $\dfrac{20x}{2x}$

Now divide this. Simply divide the 2x into both terms of the polynomial like you did above.

d) $\dfrac{8x^2 + 20x}{2x}$

Try these.

e) $\dfrac{30y^3 - 15y}{5y}$

f) $\dfrac{44a^3 - 11a^2}{-11a}$

g) $\dfrac{9x^3y^2 - 12x^2y}{3xy}$

Extending the idea to three terms is the same concept. Think of it as three problems.

h)
$$\dfrac{24x^3y^3 - 8x^2y^2 + 12xy^3}{4x\ y}$$

Divide the following. (The x can't simplify and it will need to stay in the denominator.)

i)
$$\dfrac{4}{2x}$$

That's the idea behind our last problem:

j)
$$\dfrac{50x^2 - 12x + 4}{2x}$$

Algebra I

Active Lesson: 6.6b

The idea in this section will follow something you know but which you may not have done for a while. We are going to do long division. Divide the following. If you remember how, go ahead. If not, I will walk you through it below.

a)

```
     _____
12 | 2   5   9
```

First, can 12 go into the 2. No. So, try going into the first two numbers.

```
     _____
12 | 2   5   9
 -   0
    _____
     2   5
```

Now, 12 goes into 25 two times. 2 times 12 is 24. Check for any leftovers by subtracting. There is one left over.

```
        0   2
     _____
12 | 2   5   9
 -   0
    _____
     2   5
 -   2   4
    _____
             1
```

Bring down the 9 to join the 1. 12 then goes into 19 one time.

$$
\begin{array}{r}
0\ \ 2\ \ 1 \\
12\ \overline{)\ 2\ \ 5\ \ 9} \\
-\ \ 0 \\
\overline{}\ \ 2\ \ 5 \\
-\ \ 2\ \ 4 \\
\overline{}\ \ 1\ \ 9 \\
-\ \ 1\ \ 2 \\
\overline{}\ \ 7
\end{array}
$$

You can't divide any further and so your final answer is: $21\ R7$. (Remember the R is the remainder.)
Now we are going to do it with polynomials. But it will follow the exact same process. Try this if you can. If not, I will walk you through.

b) $x\ \overline{)\ x^4\ +\ x^3\ -\ x^2\ +\ x\ +\ 1}$

We will follow the exact same steps as regular long division. How many times does x go into x^4? It can go x^3 times. x^3 times x is x^4. Check for any leftovers by subtracting. There aren't any here. Bring down the next number to continue.

$$
\begin{array}{r}
x^3 \\
x\ \overline{)\ x^4\ +\ x^3\ -\ x^2\ +\ x\ +\ 1} \\
-\ x^4 \\
\overline{}\ 0\ +\ x^3
\end{array}
$$

Play the game again. How many times does x go into x^3? It can go x^2 times. x^2 times x is x^3. Check for any leftovers by subtracting. There aren't any here. Bring down the next number to continue.

```
                x³  +  x²
         _____
    x │  x⁴  +  x³  -  x²  +  x  +  1
    -    x⁴
         _____        │
         0  +  x³           │
             -  x³          ↓
             _____
                0  -  x²
```

Play the game again. How many times does x go into $-x^2$? It can go $-x$ times. $-x$ times x is $-x^2$. Check for any leftovers by subtracting. There aren't any here. Bring down the next number to continue.

```
                x³  +  x²  -  x
         _____
    x │  x⁴  +  x³  -  x²  +  x  +  1
    -    x⁴
         _____                │
         0  +  x³                    │
             -  x³                   │
             _____            │
                0  -  x²             │
                    -  -  x²         ↓
                    _____
                       0  +  x
```

Play the game again. How many times does x go into x? It can go 1 time. x times 1 is x. Check for any leftovers by subtracting. There aren't any here. Bring down the next number to continue.

```
                x³  +  x²  -  x  +  1
         _____
    x │  x⁴  +  x³  -  x²  +  x  +  1
    -    x⁴
         _____
         0  +  x³                       │
             -  x³                       │
             _____                 │
                0  -  x²                  │
                    -  -  x²              │
                    _____          │
                       0  +  x            ↓
                          -  x
                          _____
                             0  R  1
```

You can't play again. x can't go into 1, so it is your remainder. In algebra, we will take the remainder and put it over the divisor. Here is what our answer would look like:

$$x^3 + x^2 - x + 1 + \frac{1}{x}$$

345

Try one on your own:

c) $\quad x \,\overline{\big)\, x^5 \;-\; x^4 \;+\; x^3 \;+\; x^2 \;-\; x \;+\; 2}$

Unfortunately, the problems aren't this easy. We are typically dividing by more than a simple monomial, but the plan stays the same.

$x+1 \,\overline{\big)\, 2x^4 \;+\; 5x^3 \;-\; 3x^2 \;+\; x \;+\; 6}$

Divide the first term of the binomial into the first term of the polynomial. How many times can x go into $2x^4$? It can go $2x^3$. But this time, we will multiply $2x^3$ times the binomial out front. $2x^3(x+1) = 2x^4 + 2x^3$. Check for any leftovers by subtracting. (Be careful here, you are changing the signs of both terms.) This time there are $3x^3$. Bring down the next number to continue.

$$
\begin{array}{r}
2x^3 \\
x+1 \,\overline{\big)\, 2x^4 \;+\; 5x^3 \;-\; 3x^2 \;+\; x \;+\; 6} \\
-\underline{(2x^4 \;+\; 2x^3)} \\
3x^3 \;-\; 3x^2
\end{array}
$$

Play again. How many times can x go into $3x^3$? It can go $3x^2$. Multiply $3x^2$ times the binomial out front. $3x^2(x+1)$. Check for any leftovers by subtracting. (Again, the negative sign distributes.) There are $-6x^2$. Bring down the next number to continue.

```
                2x³   +   3x²
x+1 │ 2x⁴  +  5x³  -   3x²   +  x  +  6
    -  (2x⁴  +  2x³)
                3x³   -   3x²
             -  (3x³  +   3x²)
                     -6x²   +   x
```

Play again. How many times can x go into $-6x^2$? It can go $-6x$. Multiply $-6x$ times the binomial out front. $-6x(x+1)$. Check for any leftovers by subtracting. (Again, the negative sign distributes.) There are $7x$. Bring down the next number to continue.

```
                2x³   +   3x²   -   6x
x+1 │ 2x⁴  +  5x³  -   3x²   +  x   +  6
    -  (2x⁴  +  2x³)
                3x³   -   3x²
             -  (3x³  +   3x²)
                      -
                      6x²   +   x
                   -  (6x²  -   6x)
                             7x   +  6
```

Play one more time. How many times can x go into $7x$? It can go 7. Multiply 7 times the binomial out front. $7(x+1)$. Check for any leftovers by subtracting. (Again, the negative sign distributes.) There is -1.

```
                2x³   +   3x²   -   6x   +   7
x+1 │ 2x⁴  +  5x³  -   3x²   +  x   +  6
    -  (2x⁴  +  2x³)
                3x³   -   3x²
             -  (3x³  +   3x²)
                      -
                      6x²   +   x
                   -  (6x²  -   6x)
                             7x   +  6
                          -  (7x  +  7)
                                   -1
```

Write your remainder as a fraction and your final answer is: $2x^3 + 3x^2 - 6x + 7 - \dfrac{1}{x+1}$.

Try one on your own.

d) $x-2 \overline{\smash{\big)}\, 3x^4 \ - \ 5x^3 \ - \ 4x^2 \ + \ 4x \ - \ 2}$

When you do long division, you need terms of all the degrees. For instance, the problem below is missing a term in x^3 position.

$2x+1 \overline{\smash{\big)}\, 8x^4 \ - \ 3x^2 \ - \ 5x \ + \ 4}$

Since a term is missing and we need one, we add a placeholder with a zero. From that point, the process then works the same. Finish the problem now that it has the placeholder.

e) $2x+1 \overline{\smash{\big)}\, 8x^4 \ + \ 0x^3 \ + \ 4x^2 \ - \ 5x \ + \ 4}$

Algebra I

Active Lesson: 6.7a

Simplify the following.

a) $\dfrac{x^3}{x^5}$

b) There are two ways to write your answer. One way is with the remaining x's in the denominator. The other way is with a negative exponent. Write both answers in the space below.

As we saw previously, the rules of math can't break, so the following must be true:

$$\frac{1}{x} = x^{-1}$$

Simplify the following by getting rid of the negative exponents. (Moving them to the denominator will make the exponent a positive. I've done the first one for you.)

$$y^{-7} = \frac{1}{y^7}$$

c) $a^{-3} =$

d) $x^{-12} =$

Try a real number. Simply move it to the bottom and then square it.

e) $6^{-2} =$

Or this. Move it to the bottom and cube it.

f) $2^{-3} =$

Here is another way to see what is happening with a negative exponent in the denominator.

$$\frac{1}{x^{-2}} = \frac{1}{\frac{1}{x^2}}$$

But when we divide by a fraction, we keep change flip:

$$1 \div \frac{1}{x^2} = 1 \cdot \frac{x^2}{1} = x^2$$

So, as we've seen before, negative exponents in the denominator (bottom) simply move to the numerator (top).

$$\frac{1}{x^{-2}} = x^2$$

Simplify the following by getting rid of the negative exponents. (Move them to the numerator.)

g) $\dfrac{1}{x^{-3}} =$

h) $\dfrac{1}{m^{-11}} =$

Try the following problems which involve real numbers. (When you are finished you will have removed the exponent and have a single number.)

i) $\dfrac{1}{3^{-2}} =$

j) $\dfrac{1}{2^{-4}} =$

For these next problems, notice that the entire group is raised to a negative exponent. This will cause the entire group to move.

Simplify by getting rid of the negative exponent.

k) $(5x)^{-1}$

l) $\dfrac{1}{(2x)^{-1}}$

m) $(2x)^{-2}$

n) $\dfrac{1}{(3y)^{-2}}$

In these, notice that the negative exponent doesn't include everything. Simplify by removing the negative exponents.

o) $5x^{-3}$

p) $\dfrac{1}{2y^{-5}}$

Here's another variation. Follow the rules of exponents and then simplify by removing the negative exponents.

q) $(x^3)^{-2} =$

r) $(y^{-3})^3 =$

Finally, we want to look at negative exponents and fractions.

$$\left(\frac{x}{y}\right)^{-2} = \frac{x^{-2}}{y^{-2}}$$

Below, write what you believe should be done to remove the negative exponents.

s) $\dfrac{x^{-2}}{y^{-2}} =$

Try this. The fraction will flip upside down. Then square the numerator and denominator.

t) $\left(\dfrac{2}{3}\right)^{-2}$

One more.

u) $\left(\dfrac{2a}{b}\right)^{-2}$

Algebra I

Active Lesson: 6.7b

The next lesson involves negative exponents, but it is based on ideas which we already know. Simplify the following:

a)
$$x^4 \cdot x^3 =$$

b) Which mathematical operation did you do to combine the exponents?

   a) Addition
   b) Subtraction
   c) Multiplication
   d) Division

Follow the same idea to simplify the following:

c) $x^4 \cdot x^{-3} =$

d) $y^{-3} \cdot y^8 =$

e) $a^{-4} \cdot a^{-7} =$

Previously, we learned to multiply the like terms. Simplify the following:

f) $(2x^5) \cdot (4x^3) =$

The concept is the same with negative exponents. Simplify these:

g) $(2x^5) \cdot (4x^{-3}) =$

h) $(m^5n^{-4}) \cdot (m^{-3}n^6) =$

Simplify these. Then, rewrite them without any negative exponents.

i) $(r^3s^4) \cdot (r^{-5}s^{-6}) =$

j) $(3x^{-8}y^5) \cdot (5x^{-3}y) =$

Simplify this.

k)
$$\frac{x^4}{x^2} =$$

l) What mathematical operation can you do to the exponents to allow you to simplify this?

    a) Addition
    b) Subtraction
    c) Multiplication
    d) Division

You can do the same with these:

m) $\dfrac{x^6}{x^{-2}} =$

n) $\dfrac{y^5}{y^{-4}} =$

For the next problem, recall an idea you already know.

o)
$$(x^3)^4 =$$

p) What mathematical operation did you do to simplify this?

    a) Addition
    b) Subtraction
    c) Multiplication
    d) Division

Let's extend this to negative numbers. Simplify and then rewrite without any negative exponents.

q) $(x^3)^{-4} =$

r) $(y^{-2})^6 =$

s) $(2a^{-5})^2 =$

t) $(4y^{-3})^3 =$

Algebra I

Active Lesson: 6.7c

In this activity, we are going to learn an idea called Scientific Notation. Scientific Notation is a clever system for simplifying the work of doing math on large (or small) numbers. It is based on the exponent rules we've learned and the properties of working in a system based on the number ten. Solve this exponent:

a) 10^4

Your answer should have four zeros. So, multiply the following numbers:

b) $1.7 \times 10000 =$

c) $2.9 \times 10^4 =$

d) $5.4 \times 10^4 =$

e)   Look at each of your answers. How many decimal places did each of your original numbers move?

The exponent on the 10 will simply move the number's decimal place. Try some more:

f) $2.54 \times 10^3 =$

g) $7.326 \times 10^6 =$

h) $1.89224 \times 10^7 =$

To put a number in scientific number, we must have a single number in the one's place. Then we make our exponent based on how many spots we moved the decimal place.

$$284 = 2.84 \times 10^2$$

Two spots, so 10^2. Try some.

i) $5432 =$

j) $120054 =$

When you make your number into scientific notation, zeros on the end don't need to be shown.

$$64500 = 6.45 \times 10^4$$

Try one:

k) $1250000 =$

Scientific Notation can go the other direction too. Look at the following.

$$10^{-1} = \frac{1}{10}$$

$$10^{-2} = \frac{1}{100}$$

$$10^{-3} = \frac{1}{1000}$$

So, finish the following:

l) $4.5 \times 10^{-3} = \dfrac{4.5}{1000} =$

m) $1.8 \times 10^{-2} = \dfrac{1.8}{100} =$

n) $7.835 \times 10^{-5} = \dfrac{7.835}{100000} =$

o)   Look at each of your answers. How many decimal places did each of your original numbers move?

The exponent on the 10 will move the number's decimal place. This time, because the exponent is negative, it will move left, making the number smaller. Try some more:

p) $1.3 \times 10^{-2} =$

q) $8.925 \times 10^{-6} =$

To make a small number into Scientific Notation, we again have a single number in the one's place. Then we make our exponent based on how many spots we moved the decimal place. This time the exponent will be negative.

$$.00285 = 2.85 \times 10^{-3}$$

Try some:

r)   .0556 =

s)   .0001414 =

t)   .000005 =

To take numbers out of Scientific Notation, just remember that positive exponents make the number bigger (moving the decimal to the right), and negative exponents make the number smaller (moving the decimal to the left).

Work these:

u) $5.9 \times 10^5 =$

v) $1.2950 \times 10^{-6} =$

w) $8.222 \times 10^7 =$

x) $1.879 \times 10^{-3} =$

Finally, we want to multiply and divide numbers in Scientific Notation. The idea is based on this.

Simplify the following:

y) $x^5 \cdot x^{-7} =$

z) $10^5 \cdot 10^{-7} =$

When multiplying Scientific Notation, we do the same. Just multiply the numbers out front and add the exponents on the 10's.

$$(2 \times 10^5)(3 \times 10^{-7}) = 6 \times 10^{-2}$$

Work the following.

aa) $(2 \times 10^{-2})(4 \times 10^8) =$

bb) $(3 \times 10^5)(3 \times 10^{-7}) =$

On this next problem, find the answer and then remove your answer from Scientific Notation.

$$(2 \times 10^{-3})(3 \times 10^{-2}) =$$

To understand division, work the following:

cc) $\dfrac{x^5}{x^3} =$

dd) $\dfrac{10^5}{10^3} =$

Dividing numbers in scientific notation works the problem in two parts.

$$\frac{6 \times 10^4}{3 \times 10^2} = \frac{6}{3} \times \frac{10^4}{10^2} = 2 \times 10^2 = 200$$

Divide the following. Then, remove your answer from scientific notation.

ee) $\dfrac{15 \times 10^7}{5 \times 10^2}$

ff) $\dfrac{21 \times 10^6}{7 \times 10^4}$

gg) $\dfrac{16 \times 10^4}{2 \times 10^{-2}}$

a)   What is the largest number which you could divide into 12 and 15?

Below, I have created the prime factorization for the numbers 12 and 15. Draw a circle around any shared factors.

$$12 = 2 \cdot 2 \cdot 3$$

$$15 = 5 \cdot 3$$

This shared factor is called the **GCF** (Greatest Common Factor).

b)   What is the largest number you could divide into 12 and 30?

Once again, I have created the prime factorization for the numbers. Draw a circle around any shared factors.

$$12 = 2 \cdot 2 \cdot 3$$

$$30 = 2 \cdot 3 \cdot 5$$

If you multiply the shared factors together and you have the GCF.

Create the prime factorization for the following numbers: 45, 75, and 30.

c) $45 =$

d) $75 =$

e) $30 =$

f)   Find the GCF. (Hint: A factor must now be shared across all three numbers.)

The idea is basically the same if we add in variables. Below I've made the prime factorization for $12x^2$ and $15x^3$. Circle the common factors. (Hint: x's are now factor's too.)

$$12x^2 = 2 \cdot 2 \cdot 3 \cdot x \cdot x$$

$$15x^3 = 5 \cdot 3 \cdot x \cdot x \cdot x$$

g) What is the GCF? (Be sure to multiply the x's back together.)

Next, find the GCF of $12x^2y^3$ and $30x^3y^5$

$$12x^2y^3 = 2 \cdot 2 \cdot 3 \cdot x \cdot x \cdot y \cdot y \cdot y$$

$$30x^3y^5 = 2 \cdot 3 \cdot 5 \cdot x \cdot x \cdot x \cdot y \cdot y \cdot y \cdot y \cdot y$$

h) What is the GCF?

Try some that are a bit tricky: $3(x+1)$ and $5(x+1)$. Start by making the prime factorization.

i) $3(x+1) =$

j) $5(x+1) =$

k) What is the GCF?

$8(x+3)$ and $20(x+3)$. Start by making the prime factorization.

l) $8(x+3) =$

m) $20(x+3) =$

n) What is the GCF?

As always, we can extend the idea. Put a line under each of the factors in the following:

o)
$$5(x + 2)$$

If we multiplied, we would get: $5x + 10$. But instead of distributing, suppose we wanted to go the other direction. We could pull out a GCF and we would get back to where we started.

$$5x + 10 = 5(x + 2)$$

Here's the point. If all the terms of a polynomial share a GCF, we can pull out that GCF. In essence, it is like reverse distributing. By doing so, we have turned the polynomial into factors. Factor the following by pulling out a GCF. I've done the first one for you.

$$10x^2 - 15x = 5x(x - 3)$$

p) $3y^3 + 6y^2$

q) $14x^2y^2 - 28x^2y + 21xy$

Sometimes your GCF takes everything from a term. You must leave something behind, so you put a 1. Try these:

r) $12a^3 + 8a^2 + 4a$

s) $x^2y^2 - 5x^3y^3 + 2x^4y^2$

Pull out a GCF of -1 from the following. All of the terms left behind will change signs.

t)
$$-x + 5$$

And if your first term begins with a negative, it makes the math easier to take out the negative as part of the GCF.

u) $-3y^3 + 6y^2$

v) $-2a^3 + 18a^2 - 4a$

Algebra I

Warm Up: 7.1b

Underline the two factors in the following:

a) $2(x - 2)$

b) $7(y + 3)$

Now, pull out the GCF from these:

c) $3x(x - 2) + 5(x - 2)$

d) $11y(y + 3) + 2(y + 3)$

e) $5a(a - 7) - 2(a - 7)$

f) $4m(m + 2) - (m + 2)$

In this activity, we are going to learn a factoring method called grouping. I will walk you through the idea here. Pull out a GCF from the following:

g) $x^2 + 2x$

h) $3x + 6$

Now I've combined everything into one polynomial. Pull out a GCF from the first two terms and pull out a GCF from the last two terms.

i)
$$x^2 + 2x + 3x + 6$$

j) Although it looks strange, your new polynomial has a GCF of $(x + 2)$. In the space below, show how you could pull it out to get the following: $(x + 2)(x + 3)$.

Let's do it again. Pull out a GCF from the following:

k) $y^2 + 4y$

l) $3y + 12$

Again, I've combined everything into one polynomial. Pull out a GCF from the first two terms and pull

m) out a GCF from the last two terms.

$$y^2 + 4y + 3y + 12$$

n) This rearranged polynomial now has a GCF of $(y + 4)$. In the space below, show how you could pull it out to get the following: $(y + 4)(y + 3)$.

Try some on your own. Always pull out the GCF of the first two terms and then of the last two terms. If you have done it correctly, you will have a matching binomial which then becomes a GCF.

o) $x^2 + 5x + 3x + 15$

p) $x^2 - 6x + 4x - 24$

In this one, the back two terms start with a negative, so pull out a negative GCF.

q) $y^2 + 2y - 6y - 12$

Here, remember, if you pull out a GCF, you can't leave a spot empty. You put a 1 in its place.

r) $x^2 + x + 6x + 6$

369

Algebra I

Active Lesson: 7.2a

Below, I have multiplied out a binomial. (F.O.I.L.: First Outside Inside Last).

$$(x + 5)(x + 6) = x^2 + 5x + 6x + 30$$

a) Combine the like terms and write out the final trinomial.

Here I have multiplied another.

$$(y - 7)(y - 2) = y^2 - 7y - 2y + 14$$

b) Combine the like terms and write out the final trinomial.

Multiply out this one on your own. Be sure to combine the like terms.

c) $(x - 3)(x + 5) =$

Factoring polynomials is nothing more than recognizing how we multiplied and then going backward.

$$x^2 \boxed{+ 5x + 6x} + 30$$
$$\wedge$$
$$5*6$$

1. What two numbers multiplied to get the 30?
2. And add to get $11x$?

$$y^2 \boxed{- 7y - 2y} + 14$$
$$\wedge$$
$$-7*-2$$

1. What two numbers multiplied to get the 14?
2. And add to get -9y?

371

Factor the following: (What two numbers did we multiply to get the last number and add to get the middle?)

d) $x^2 + 10x + 16 = (x+\_\_)(x+\_\_)$

e) $x^2 + 10x + 25 = (x+\_\_)(x+\_\_)$

f) $y^2 - 7y + 12 = (y-\_\_)(y-\_\_)$

On this next one, remember that the signs of the numbers matter. What two numbers multiply to get -18 and add to get -3? One number will need to be positive and one will need to be negative.

g)
$$y^2 - 3y - 18 = (y-\_\_)(y+\_\_)$$

Try these on your own.

h) $x^2 + 12x + 20 =$

i) $x^2 - 12x + 35 =$

j) $x^2 + 6x - 7 =$

Algebra I

Active Lesson: 7.2b

In this activity, we are once again going to factor trinomials. These are going to look quite difficult because they involve two variables. However, I want you to see that it doesn't change much of anything. As I'm sure you will figure out, I'm using the same problems from the last activity. The only change is the extra variable. The point is to show you that the process is nearly identical. However, work them without turning back to your previous answers.

Below, I have multiplied out a binomial. (F.O.I.L.: First Outside Inside Last).

$$(x + 5y)(x + 6y) = x^2 + 5xy + 6xy + 30y^2$$

a) Combine the like terms and write out the final trinomial.

Here I have multiplied another.

$$(y - 7x)(y - 2x) = y^2 - 7xy - 2xy + 14x^2$$

b) Combine the like terms and write out the final trinomial.

Multiply out this one on your own. Be sure to combine the like terms.

c) $(x - 3y)(x + 5y) =$

To factor these trinomials, we are still playing the exact same game. What two numbers multiply to get the last term and add to get the middle? Having the extra variable on the end doesn't change anything about the process.

Factor the following:

d) $$x^2 + 10xy + 16y^2 = (x + \underline{\quad} y)(x + \underline{\quad} y)$$

e) $$x^2 + 10xy + 25y^2 = (x + \underline{\quad} y)(x + \underline{\quad} y)$$

373

f)
$$y^2 - 7xy + 12x^2 = (y-\underline{}x)(y-\underline{}x)$$

As we saw before, remember that the signs of the numbers matter. What two numbers multiply to get -18 and add to get -3? One number will need to be positive and one will need to be negative.

g)
$$y^2 - 3xy - 18x^2 = (y-\underline{}x)(y+\underline{}x)$$

Try these on your own.

h)
$$x^2 + 12xy + 20y^2 =$$

i)
$$x^2 - 12xy + 35y^2 =$$

j)
$$x^2 + 6xy - 7y^2 =$$

Algebra I

Active Lesson: 7.3a

Factor the following:

a) $x^2 - 2x - 15$

In the last two activities, we learned how to factor trinomials like the one above. But it isn't always so easy. So far, we've done problems which always have a 1 at the beginning. Changing that causes a significant challenge. But, like before, the idea is still rooted in doing the multiplication and finding a pattern. I've factored each of the following. Find the connection between the underlined numbers.

$$\underline{6}x^2 + 19x + 10 = (\underline{2}x + 5)(\underline{3}x + 2)$$

$$\underline{9}x^2 + 9x + 2 = (\underline{3}x + 1)(\underline{3}x + 2)$$

$$\underline{8}x^2 + 10x - 3 = (\underline{4}x - 1)(\underline{2}x + 3)$$

$$\underline{5}x^2 - 3x - 2 = (\underline{5}x + 2)(\underline{x} - 1)$$

b)   What is the connection between the underlined numbers?

375

There are two methods for solving a trinomial with a non-zero in front. In a classroom, I no longer teach the method we will do in this lesson because the alternative is so much easier. However, it works, and if you have strong math skills, you might find this to be faster.

The method is called "Trial and Error," and if that doesn't sound too promising, there is a reason for that. Literally, you must guess at the possible factors of the first term and guess at the possible factors of the last term. Then, you multiply them out to see if you were correct.

Below, I've listed all the possible factors. Multiply and find which one must be correct. Circle the answer.

c)

$$3x^2 + 19x + 6$$

$$(3x + 6)(x + 1)$$

$$(3x + 1)(x + 6)$$

$$(3x + 2)(x + 3)$$

$$(3x + 3)(x + 2)$$

d)

$$2x^2 - 10x + 12$$

$$(2x - 12)(x - 1)$$

$$(2x - 1)(x - 12)$$

$$(2x - 6)(x - 2)$$

$$(2x - 2)(x - 6)$$

$$(2x - 4)(x - 3)$$

$$(2x - 3)(x - 4)$$

e)

$$6x^2 + 10x + 4$$

$$(6x + 4)(x + 1)$$

$$(6x + 1)(x + 4)$$

$$(3x + 1)(2x + 4)$$

$$(3x + 4)(2x + 1)$$

$$(3x + 2)(2x + 2)$$

f)

$$12x^2 - 6x - 6$$

$$(12x - 6)(x + 1)$$

$$(12x + 6)(x - 1)$$

$$(6x - 6)(2x + 1)$$

$$(6x + 1)(2x - 6)$$

$$(4x - 6)(3x + 1)$$

$$(4x + 1)(3x - 6)$$

$$(12x - 2)(x + 3)$$

$$(12x + 3)(x - 2)$$

$$(6x - 2)(2x + 3)$$

$$(6x + 3)(2x - 2)$$

$$(4x - 3)(3x + 2)$$

$$(4x + 2)(3x - 3)$$

It is obviously an unpleasant approach, but it does get easier with practice. In our next lesson, we'll learn a much less cumbersome method. For now, try a couple on your own. One easier problem and one harder.

g) $2x^2 + 1x - 10$

h) $6x^2 - 11x + 4$

Algebra I

Active Lesson: 7.3b

Below, I'm multiplying out a pair of binomials.

$$(x + 2)(x + 3)$$

$$x^2 + 2x + 3x + 6$$

$$x^2 + 5x + 6$$

Now, I've written it in the opposite direction, going from the trinomial to the factors.

$$x^2 + 5x + 6$$

$$x^2 + 2x + 3x + 6$$

$$(x + 2)(x + 3)$$

a) Do the same thing. First foil it out, showing all the terms. Then, combine like terms.

$$(x + 4)(x + 2)$$

b) Now, write the same problem in the opposite direction, going from the trinomial to the factors.

Factoring is un-foiling a trinomial. However, going backward should include that middle step, but we haven't. Yet, the method we are going to learn in this activity, will. It is called the a/c method. Let me show you where the middle step is found, and how we get from the middle step to the final answer.

Let's look again at: $x^2 + 5x + 6$.

The a term is the number in front of x^2 and the c term is the last number. Multiply them together.

$$x^2 + 5x + 6$$

In this problem we get 6. Now, play the same game we did before. What two numbers multiply to get 6 and add to get the 5 in the middle. The answer is 3 and 2. Below, I've split the $5x$ into these two numbers. I haven't done anything wrong. They still add to 5x. Now, factor by grouping. Take the GCF of the first two terms and the GCF of the last two terms.

c) $$x^2 + 2x + 3x + 6$$

d)   Now, you should have a matching term $(x + 2)$. Pull that out as a GCF.

You now have the factored trinomial. We skip this process when the first term of the trinomial is a 1, because it makes things more time consuming. But, if we are really "un-FOIL-ing" a trinomial, we should include the entire process. Try one yourself:

- Multiplying $1 \cdot 8 = 8$.
- What two terms multiply to get 8 and add to get 6?
- Make those your middle terms.
- Factor by grouping.

e) $$x^2 + 6x + 8$$

When the first term of a trinomial is not a 1, this is the approach which will always work. The a/c method will take the entire trinomial apart one step at a time and ensure a solution. Let's try one:

$$6x^2 + 10x + 4$$

- We multiply $a \cdot c$ and get 24.
- What two numbers multiply to get 24 and add to get 10? 6 and 4.
  We make those the new middle term: $6x^2 + 6x + 4x + 4$
- Factor by grouping to complete the problem.

f) $$6x^2 + 6x + 4x + 4$$

Again, this method will always work, and so generally saves a great deal of time over the Trial-and-Error method.

Try some more:

g) $2x^2 - 10x + 12$

h) $2x^2 + 1x - 10$

In this next problem, when you go to factor by grouping, it will seem like there is nothing to take from one of the groups. Remember, in those situations, you must take something and so you take a 1.

i) $3x^2 + 19x + 6$

In this next one, when you factor by grouping, the second group will start with a negative. Remember, in those situations, you take the negative out as part of your GCF. I've started it for you. Finish it by factoring by grouping.

$$12x^2 - 6x - 6$$
$$12x^2 + 6x - 12x - 6$$

j)

Try another on your own:

k) $8x^2 + 10x - 3$

Algebra I

Active Lesson: 7.4a

I have multiplied out these binomials. Finish by combining like terms.

a)
$$(x - 5)(x - 5) = x^2 - 5x - 5x + 25$$

b)
$$(y + 3)(y + 3) = y^2 + 3y + 3y + 9$$

Multiply the following binomials:

c) $(y - 9)(y - 9)$

d) $(m + 7)(m + 7)$

e) When you are done multiplying this type, you always get a trinomial. What is special about the middle of these trinomials?

f) What causes you to create a middle term like this?

g) When you are finished multiplying, you are always left with a trinomial. What is the sign of the third (final) term of the trinomial?

h) There is also something that determines the sign of the middle term. What is it?

As we've seen, factoring is simply following a pattern to un-do a polynomial. You could do the same process that we've done before and you will get the answer. But there is a shortcut which saves so much time, it is worth learning. Look at this trinomial.

$$y^2 + 6y + 9$$

Check three things:

1) Does it have a square-root on the first term? Yes. y.
2) Does it have a square-root on the back term? Yes. 3.
3) Is the middle term double those two square roots multiplied together? Yes. $2 \cdot 3 \cdot y = 6y$.

Then, to factor it, write the square roots as two binomials and give them each the sign of the original middle term.

$$(y + 3)(y + 3)$$

Try one:

$$x^2 - 8x + 16$$

1) Does it have a square-root on the first term?
2) Does it have a square-root on the back term?
3) Is the middle term double those two square roots multiplied together?

i)   If you answered yes to all of those questions, factor it. (Remember, both terms get the sign of the original middle term.)

Work some more.

j) $x^2 + 10x + 25$

k) $x^2 - 22x + 121$

This next type has an a term, but it doesn't change anything. If it still has a square root, the process remains the same:

l) $9x^2 + 12x + 4$

m) $25x^2 - 40x + 16$

These problems have two variables, but the process remains the same.

n) $x^2 + 10xy + 25y^2$

o) $16x^2 - 24xy + 9y^2$

With so much to factoring, I generally avoid tricky problems. But let me show you one you may see in a textbook. The following polynomial doesn't fit the pattern and so can't be factored.

$$9a^2 + 20x + 16$$

p) Why doesn't it fit?

q) Let's look at one final idea. Write the following number in terms of its prime factors:

$$30 =$$

Notice that there are three factors.

The first step in factoring is always to pull out a GCF. Every time. Sometimes, it may not appear as if you see a factoring pattern, but it may be that you've missed the GCF.

$$4x^2y + 40xy + 100y$$

This has a GCF of $4y$, so pull it out:

$$4y(x^2 + 10x + 25)$$

Then you can finish factoring. But, remember, a GCF is a factor too! So, your answer should include three factors.

$$4y(x + 5)(x + 5)$$

Try some on your own. (First, pull out the GCF.)

r) $x^2y + 6xy + 9y$

s) $2a^2b^3 + 36ab^3 + 162b^3$

Algebra I

Active Lesson: 7.4b

Let's learn another factoring shortcut by recalling a pattern. I have multiplied out these binomials. Finish by combining like terms.

a)
$$(x + 5)(x - 5) = x^2 + 5x - 5x - 25$$

Multiply the following binomials:

b) $(y - 9)(y + 9)$

c) $(m - 7)(m + 7)$

d) What keeps happening to the middle?

e) Why does this keep happening?

f) When you are finished multiplying, you are always left with a binomial. What is the sign of this binomial?

Factoring is recognizing the pattern of taking a polynomial in reverse. Here's what is needed for our newest shortcut.

$$x^2 - 9$$

1) Does the first term have a square root? Yes. x.
2) Does the last term have a square root? Yes. 3.
3) Is the sign in-between a negative? Yes.

Then, to factor it, write the square roots as two binomials and give them each a different sign.

$$(x - 3)(x + 3)$$

Try on your own.

g) $$y^2 - 16$$

1) Does the first term have a square root?
2) Does the last term have a square root?
3) Is the sign in-between a negative?

If you answered yes to all of those questions, factor it. (Remember, both terms get opposite signs.)

Work some more.

h) $x^2 - 25$

i) $x^2 - 121$

This next type has an a term, but it doesn't change anything. If it still has a square root, the process remains the same:

j) $9x^2 - 4$

k) $25x^2 - 16$

These problems have two variables, but the process remains the same.

l) $x^2 - 25y^2$

m) $16x^2 - 9$

These next two reverse the order, but that's okay. Just reverse the way you write it.

$25 - x^2 = (5 - x)(5 + x)$

n)   Factor: $121 - y^2$

o)   Can you give me the prime factorization of 7? Why or why not?

Multiply out each of the following:

p) $(x + 3)(x + 3)$

q) $(x - 3)(x - 3)$

r) $(x + 3)(x - 3)$

s) Notice that we've done each possible sign combination. Do any of these multiply out to get $x^2 + 9$?

If you have a binomial with square roots on the front and square roots on the back, you must have a negative in the middle. Binomials with a positive in the middle can't be factored. There is no combination that can multiply to create it. So, $x^2 + 9$ is a prime number.

t) Give me the prime factorization of 50.

$$50 =$$

This next problem has a higher power. But it still fits the pattern:

$$x^4 - 16$$

1) Does the first term have a square root? Yes. x^2
2) Does the last term have a square root? Yes. 4.
3) Is the sign in between a negative? Yes.

So, we factor and get:$(x^2 - 4)(x^2 + 4)$. But just like when we find a prime factorization, sometimes the numbers can go further. $(x^2 + 4)$ is prime. But $(x^2 - 4)$ can be factored further. In the space below,
u) factor $x^4 - 16$ ad far as it will go. Remember, $(x^2 + 4)$ is a factor and must be part of your final answer.

Try another. Factor as far as it will go.

v) $a^4 - b^4$

Finally, as we saw in the last section, pulling out a GCF is always the first step in factoring. These next polynomials fit our pattern (called a perfect square binomial) but you must start by pulling out the GCF. Remember to put the GCF as part of the final factored solution.

w) $x^2y - 100y$

x) $5a^2b^3 - 125b^3$

y) $2x^4 - 32$

In this activity, we will see that there is also a pattern for binomials which have cubes on the front and back.

Finish multiplying the problem which I have started below. I'd recommend foiling the last two binomials. Then you will need to multiply your answer by the remaining $(x - 3)$.

a)
$$(x - 3)(x - 3)(x - 3)$$

If you did this correctly, you should have gotten a very complicated answer. Here's the point, look at this binomial:

$$x^3 - 27$$

It has cube roots on the first term and on the last term, but it doesn't turn into this:

$$x^3 - 27 \neq (x - 3)(x - 3)(x - 3)$$

The reality is a bit more unfortunate. To factor a binomial which has cube roots on the first term and the last term, you need to know two things:

1) The pattern $(a\ b)(a^2\ ab\ b^2)$.
2) And the acronym: S.O.A.P.

Let's walk through a problem.

$$x^3 - 27$$

1) Does it have a cube root on the front? Yes. x.
2) Does it have a cube root on the back? Yes. 3.

Note: These can have a positive or a negative in between!

Now, we will fit this to our pattern $(a\ b)(a^2\ ab\ b^2)$. a equals the cube root of the first term. b equals the cube root of the last term.

So:

$$a = x$$
$$b = 3$$

Then we put those into our pattern:

$a = x$

$b = 3$

$a^2 = x^2$

$ab = 3x$

$b^2 = 9$

$$(x\ 3)(x^2\ 3x\ 9)$$

But we also need signs between the terms. That's where S.O.A.P. comes in. The acronym stands for:

S = Same

O = Opposite

A = Always

P = Positive

So, first we put in the same sign from the original problem: $x^3 - 27$ The sign was negative, so our first sign is negative.

$$(x - 3)(x^2\ 3x\ 9)$$

Opposite means the next sign is always the opposite of the first.

$$(x - 3)(x^2 + 3x\ 9)$$

And Always Positive means that the last sign is always a positive.

$$(x - 3)(x^2 + 3x + 9)$$

Try one.

b)
$$x^3 - 8$$

Try another. Notice that the sign in-between is a positive this time.

c)
$$y^3 + 64$$

Nothing changes if both the terms are variables:
$$x^3 - y^3$$

$a = x$

$b = y$

$a^2 = x^2$

$ab = xy$

$b^2 = y^2$

d) Finish this problem.

This next problem is more complicated, but the process is still the same.
$$27r^3 - 8s^3$$

$a = 3r$

$b = 2s$

$a^2 = 9r^2$ (Notice that all of a has been squared.)

$ab = 6rs$ (Notice that all of a and b have been multiplied.)

$b^2 = 4s^2$ (Notice that all of b has been squared.)

e) Finish this problem.

Finally, if there is a GCF, always pull it out first.

$$2x^3z + 54y^3z$$

f)   Factor. Be sure to include the GCF as part of the answer.

Algebra I

Active Lesson: 7.5

Factoring gets more difficult when all the different patterns are mixed together. When we are presented with that situation, there is a course of action we should follow:

1) Check first for a Greatest Common Factor.
2) Check for square roots on the front and back. If you have them, you may be dealing with:
   a. A perfect square trinomial.
   b. A perfect square binomial.
3) Check for cube roots on the front and back. You may have a difference of cubes.
4) If there are four terms, try factoring by grouping.
5) If none of those apply, you have a traditional factoring problem:
   a. If the leading coefficient is 1, you can find two numbers to multiply to get the last term and add to get the middle term.
   b. If the leading coefficient isn't 1, use the a/c method.

Use the approach above to factor the following:

a) $x^2 - 2x - 35$

b) $3x^2 + 16x - 12$

c) $x^2 - 8x + 16$

d) $x^2 - 9$

e) $x^2y - 15xy + 56y$

f) $x^3 - 27y^3$

g) $16x^4 + 16x^3 + 4x^2$

h) $x^2 - 6x - 27$

i) $2x^2 + 11x + 15$

j) $x^4 - 81$

k) $x^5 + x^3y^2$

l) $x^2 + x + 3x + 3$

Multiply each of the following:

a) $7 \cdot 0$ b) $0 \cdot 7$ c) $x \cdot 0$ d) $0 \cdot x$

e) $(x - 2) \cdot 0$ f) $0 \cdot (x - 2)$

g) What is the point that I'm trying to make about multiplying by zero? (Don't overthink it.)

Suppose that I have the following puzzle:

$$x \cdot y = 0$$

If x turns out to be zero then the entire equation would be zero. Likewise, if y turns out to be zero then the entire equation would be zero. So, to solve this equation, there must be two answers, because either could be true.

$$x = 0 \text{ or } y = 0$$

The same idea carries over to polynomials. I've factored the equation below.

$$x^2 - 7x - 8 = 0$$

$$(x - 8)(x + 1) = 0$$

If either of the factors turned out to be zero, it would make the equation equal zero. So, there must be two answers, because either could be true. But we want to know the value of x, therefore we have a little more work to do. Solve each of the following equations for x.

h) $(x - 8) = 0$ i) $(x + 1) = 0$

j)   There are two values of x which make the equation zero. Below, put those two values of x into the trinomial $x^2 - 7x - 8$ and confirm that they make it zero.

Try one on your own. Solve the following equation for y. You should get two answers.

k) $y^2 - 10y + 16 = 0$

This idea is called the zero-product property, and to use it, we must have a zero on one side of the equation. Solve the following, by moving everything to one side of the equation so that the other side is zero. Then, with the idea above, factor and find the two solutions. I've started the first one for you.

l) $x^2 = 81$

$x^2 - 81 = 0$

m) $3t^2 = -13t - 4$

On this next problem, send the x^2 to the right side. This makes it positive and makes the factoring easier. It doesn't matter if you make the room on the right or the left equal to zero; it will work either way.

n) $-x^2 = 2x - 63$

On this problem, you must multiply out the left side and then bring over the $2x$.

o) $(x + 2)(x + 3) = -2x$

Before we end this activity, something odd can happen when you pull out a GCF.

$$2x^2 - 32 = 0$$

$$2(x^2 - 16) = 0$$

$$2(x - 4)(x + 4) = 0$$

Now, there are three factors being multiplied and they all equal zero. So, below, I've set them all equal to zero.

$$2 = 0 \text{ or } (x - 4) = 0 \text{ or } (x + 4) = 0$$

But two can't equal zero. Therefore, it won't offer us a solution. However, what if I started with the this:

$$2x^3 - 32x = 0$$

$$2x(x^2 - 16) = 0$$

$$2x(x - 4)(x + 4) = 0$$

Here, the GCF had an x in it. This could make a third solution.

$$2x = 0 \text{ or } (x - 4) = 0 \text{ or } (x + 4) = 0$$

$$x = 0 \text{ or } x = 4 \text{ or } x = -4$$

Try the following:

p) $2x^2 - 4x - 48 = 0$ q) $2x^3 + 7x^2 + 3x = 0$

401

Algebra I

Active Lesson: 7.6b

In this activity, we will need our skills working with quadratic equations to solve word problems. First, we will return to the ideas of consecutive integers.

The product of two consecutive integers is 210. Find the integers.

Here is our dictionary. Recall that two consecutive numbers would look like this:

Math	English
x	1st Number
x+1	2nd Number

Previously, we would add the consecutive integers. This time, we will multiply. So, our equation would be:

$$x(x + 1) = 210$$

Now, distribute.

$$x^2 + x = 210$$

As we learned before, we need the right side to equal zero, so we move the 210 to the other side.

$$x^2 + x - 210 = 0$$

Factoring we get:

$$(x + 15)(x - 14) = 0$$

Then, using the zero product property:

$$x + 15 = 0 \ or \ x - 14 = 0$$

$$x = -15 \ or \ x = 14$$

There are actually two possible consecutive integers that would make this possible.

$$x = -15, -14 \ or \ x = 14, 15$$

Try a problem on your own.

a)   The product of two consecutive integers is 156. Find the integers.

The idea can be extended to problems involving the area of a rectangle.

The area of a backyard is 224 ft^2. The length of the backyard is 2 feet more than the width. Find the length and the width of the backyard.

Math	English
x	Length
x+2	Width

The area of a rectangle is:

$$A = L \cdot W$$

$$224 = x(x + 2)$$

b) Finish solving the equation. This time, it is the left side which we need to become zero. However, that doesn't change the process.

c) What are the missing length and width?

Let's look at one more type of problem.

A right triangle has a hypotenuse of 13. One leg is 7 more than the other leg. Find the length of both legs.

Our dictionary would be:

Math	English
x	Leg One
x+7	Leg Two
13	Hypotenuse

This problem is relying on the Pythagorean theorem.

$$a^2 + b^2 = c^2$$

Substituting from our dictionary, we get:

$$(x)^2 + (x+7)^2 = 13^2$$

Cleaning this equation is up difficult. $(x+7)^2$ is $(x+7)(x+7)$ and must be foiled out.

$$x^2 + x^2 + 14x + 49 = 169$$

Next:

$$2x^2 + 14x + 49 = 169$$

$$2x^2 + 14x - 120 = 0$$

To solve, the first step would be to factor out a 2.

$$2(x^2 + 7x - 60) = 0$$

d) Factor and use the zero product property to solve for x. (Hint: 2 will never equal zero, so it has nothing to do with the answer. You will also get a negative number, which is impossible for the length of a side of a triangle. That answer will be discarded.)

Work this final problem on your own.

e) A right triangle has a hypotenuse of 10. One length of one leg is 2 more than the other. Find the legs of the right triangle.

405

In the math world, something is wrong with the following expression:

$$\frac{81}{0}$$

a) Explain the problem.

Look at these three expressions. What values of x would cause the same problem to occur?

b) $\dfrac{4}{x}$

c) $\dfrac{7}{x-2}$

d) $\dfrac{102}{5+x}$

This next problem has two values of x which would cause the problem. Factor the denominator in order to find those two values.

e) $\dfrac{1}{x^2 + 9x + 18}$

Values which make a fraction undefined are called restricted values. They aren't allowed. When looking for restricted values, the top (numerator) has nothing to do with the issue. (It can't cause the expression to be undefined.) In this next problem, find the restricted values. There are two, and disregard the numerator.

f) $\dfrac{x^2 + 5x - 7}{x^2 + 5x - 14}$

In this activity, we have begun working with something called a rational expression. A rational expression is a fraction with a polynomial on top (numerator) and a polynomial on bottom (denominator). We've already seen that certain values are restricted because they could make the rational expression undefined. But the variable in a rational expression is permitted to take on any other value.

When we are interested in knowing the outcome of a rational expression for an acceptable value it is called evaluating the rational expression. As we've seen before, it just means that we are substituting a value in for the variable.

Evaluate the following rational expression when $x = 2$.

$$\frac{x+3}{x+4}$$

Just substitute the value in. Remember, when you replace a variable with a value put it in a parenthesis.

$$\frac{(2)+3}{(2)+4} = \frac{5}{6}$$

We use the parenthesis so that we don't miss out on any multiplication.

Evaluate the following rational expression when $x = -1$.

$$\frac{2x^2+5}{x-4}$$

When you evaluate, be sure to follow the order of operations.

$$\frac{2(-1)^2+5}{(-1)-4} = \frac{2(1)+5}{-1-4} = \frac{7}{-5} = -\frac{7}{5}$$

Try one. Evaluate the following rational expression when $x = -2$.

g) $\dfrac{2x^2+5}{x-4}$

Work these. There are three parts to each problem. (There may be answers which are fractions and that is fine.)

Evaluate the following rational expression when:

h)

$$\frac{x^2-5}{x+1}$$

   a) $x = 5$
   b) $x = -3$
   c) $x = 0$

Evaluate the following rational expression when:

i)

$$\frac{2x^2 + 3x - 4}{x^2 - 1}$$

a) $x = 1$

b) $x = -4$

c) $x = -2$

Look at this fraction.

$$\frac{6}{8}$$

Typically, we reduce it in our head and know that the answer is $\frac{3}{4}$. But what we are really doing is making the prime factorization.

$$\frac{2 \cdot 3}{2 \cdot 2 \cdot 2}$$

You can cancel a two in the numerator and in the denominator.

$$\frac{2 \cdot 3}{2 \cdot 2 \cdot 2}$$

And this is how we get $\frac{3}{4}$.

Below, I've taken another fraction $\frac{42}{70}$ and done the prime factorization for the numerator and the denominator.

$$\frac{42}{70} = \frac{2 \cdot 3 \cdot 7}{2 \cdot 5 \cdot 7}$$

a)   Reduce the matching factors for $\frac{42}{70}$. What is the reduced fraction?

I've now factored an algebraic expression called a Rational Expression. Cancel any matches to reduce it.

$$\frac{42x^2y^5}{70x^3y^3} = \frac{2 \cdot 3 \cdot 7 \cdot x \cdot x \cdot y \cdot y \cdot y \cdot y \cdot y}{2 \cdot 5 \cdot 7 \cdot x \cdot x \cdot x \cdot y \cdot y \cdot y}$$

b)   What is the final reduced expression? (Put any leftover x's and y's back together. Be careful, leftovers on the top, should be written on the top. Leftovers on the bottom should be written on the bottom.)

One of the primary reasons we worked on factoring is to be able to do tricks like the following.

$$\frac{3x - 21}{x^2 - 49} = \frac{3(x - 7)}{(x - 7)(x + 7)}$$

A group in parenthesis is just a number. So, if the top and bottom have any matching parenthesis, the group can be cancelled. Cancel the matches in the expression. What is the final fraction?

c)

$$\frac{3(x - 7)}{(x - 7)(x + 7)}$$

Try one on your own. Remember, only factors can be cancelled. You can't cancel until you have turned it into factors.

d) $$\frac{5z + 25}{z^2 + 3z - 10}$$

Your answer should be $\frac{5}{z-2}$.

Try some more.

e) $$\frac{x^2 - 4}{x^2 + 3x - 10}$$

f) $$\frac{2x^2 + 3x - 9}{x^3 - 9x}$$

g) $\dfrac{2x^2 + 20x + 50}{2x^2 + 4x - 30}$

h) $\dfrac{3x^2 + 3x - 60}{6x^2 - 96}$

i) $\dfrac{m^3 + 27}{2m^2 - 18}$

Algebra I

Warm Up: 8.1c

Our next idea is the basis for a very simple, but important, math trick.

Multiply the following:

a)
$$-1(6 - x)$$

Circle the choice below which, mathematically, is the same as your answer.

b) $-6 - x$ $x + 6$ $x - 6$

Now, factor out a -1 from this expression:

c)
$$(2 - x)$$

Circle the choice below which, mathematically, is the same as your answer.

d) $-1(x - 2)$ $-1(-x - 2)$ $-1(x + 2)$

Suppose we start with a binomial that has a negative in-between the terms. We can turn the binomial around by pulling out a negative one. Why is this important? Because we will often come across factors that could cancel if only they were reversed.

$$\frac{(2 - x)}{(x - 2)}$$

Pull the negative one out and we can fix this. (A negative one and a negative are really the same thing. So, I just use the negative sign, and I put it out in front of the fraction. In a later section, you will see why this is a good strategy.)

$$-\frac{(x-2)}{(x-2)}$$

Now, the factors can cancel and you get:

$$-\frac{1}{1} = -1$$

Try one: (It doesn't matter if you want to turn around the binomial on the top or bottom. It works the same. Just put the negative out front.)

e) $\dfrac{(x-6)}{(6-x)}$

Obviously, the problems get more complicated, but the idea remains the same. Anytime you want to turn around a term, put a negative in front of the fraction. Try these: (Factor first and it will reveal if you need to turn something around.)

f) $\dfrac{15-5x}{x^2-9}$

g) $\dfrac{x^2-x-30}{36-x^2}$

h) $\dfrac{1-x^2}{x^3-11x^2+10x}$

Algebra I

Active Lesson: Section 8.2a

In this activity, we will learn to multiply rational expressions. Algebraic fractions are still just fractions, and so they follow all of the same rules. Multiply these fractions. (Top times top. Bottom times bottom.)

a)
$$\frac{1}{3} \cdot \frac{2}{5} =$$

But if the fractions can be reduced, we cancel any factors on the top with any match on the bottom. Here I've shown the prime factorization.

$$\frac{6}{5} \cdot \frac{10}{8} = \frac{2 \cdot 3}{5} \cdot \frac{2 \cdot 5}{2 \cdot 2 \cdot 2}$$

Matching factors are cancelled.

$$\frac{6}{5} \cdot \frac{10}{8} = \frac{2 \cdot 3}{\cancel{5}} \cdot \frac{\cancel{2} \cdot \cancel{5}}{2 \cdot \cancel{2} \cdot 2} = \frac{3}{2}$$

The process with rational expressions is identical. Factor everything on the top and factor everything on the bottom.

$$\frac{10xy}{6y^2} \cdot \frac{2xy}{8x^2} = \frac{2 \cdot 5 \cdot x \cdot y}{2 \cdot 3 \cdot y \cdot y} \cdot \frac{2 \cdot x \cdot y}{2 \cdot 2 \cdot 2 \cdot x \cdot x}$$

Now, anything on the top can cancel with any match on the bottom. Make the cancellations in the space below.

$$\frac{2 \cdot 5 \cdot x \cdot y}{2 \cdot 3 \cdot y \cdot y} \cdot \frac{2 \cdot x \cdot y}{2 \cdot 2 \cdot 2 \cdot x \cdot x}$$

b) Put everything remaining into one fraction. (Just slide everything together.) Multiply the factors on the top back together, and multiply the factors on the bottom back together. (Remember, leftovers on the top must stay on the top. Leftovers on the bottom must stay on the bottom.) Write your answer in the space below.

Now look at the following problem:

$$\frac{4x-8}{x+6} \cdot \frac{x^2+6x}{x^2-4}$$

The most common mistake students will make here is that they will try to cancel parts of numbers. For instance, you cannot cancel the x^2's on the top and bottom. Because of the addition and subtraction, the x^2's are part of a more complex number. It would be like trying to cancel the ones in the fraction below. You can't.

$$\frac{17}{135}$$

The ones are digits, not factors. In a similar way, it is like the x^2's are digits. They are part of a number. They aren't factored. Below, I've factored the problem. Factors can be cancelled.

$$\frac{4x-8}{x+6} \cdot \frac{x^2+6x}{x^2-4} = \frac{4(x-2)}{(x+6)} \cdot \frac{x(x+6)}{(x-2)(x+2)}$$

Notice that there are terms that involve addition and subtraction, but they are factors. If the entire
c) factor matches, it is a matching number. In the space below, cancel any matching factors between the top and bottom.

$$\frac{4(x-2)}{(x+6)} \cdot \frac{x(x+6)}{(x-2)(x+2)}$$

Next, we make one fraction. Any factors on the top remain on top. Any factors on the bottom remain on the bottom. (We could F.O.I.L. out the binomials, but in problems like this, it is preferred to just leave
d) them as factors.) In the space below, make one fraction.

Try one on your own. Remember, factor everything. Cancel any matches. And, then, make one fraction out of what remains. (We leave binomials unfactored, but multiply single terms back together. In this, you will have a 3x and a 4 left on top. Multiply those: $3x \cdot 4 = 12x$.)

e) $\dfrac{3x^2+9x}{x+1} \cdot \dfrac{4x+4}{x^2-9}$

Work some more.

f) $\dfrac{2x-4}{6x} \cdot \dfrac{4x^2}{8x-16}$

g) $\dfrac{4x-6}{2x+10} \cdot \dfrac{x^2+3x-10}{2x^2-x-6}$

h) $\dfrac{5x+15}{x^2+8x+16} \cdot \dfrac{x^2-16}{10x+10}$

In this final problem, you need the trick where we turn a binomial around. Remember, you bring a negative sign out front of the fraction. At the end of the problem, simply bring that negative sign over as part of your final answer.

i) $\dfrac{6x-6}{15-3x} \cdot \dfrac{x^2-25}{3x^2-5x+2}$

Algebra I

Active Lesson: 8.2b

In this activity, we will look at dividing rational expressions. As always, it will follow the standard rules of math. We are still just dividing fractions. So, we just need to remember: Keep, Change, Flip.

$$\frac{5}{7} \div \frac{15}{28}$$

Keep the first fraction as it is. Change the sign to multiplication. Then, flip the last fraction upside down.

$$\frac{5}{7} \cdot \frac{28}{15}$$

Next, we do exactly as before. I've factored everything below. Cancel any matches on the top and bottom and write the leftovers as one fraction.

a)
$$\frac{5}{7} \cdot \frac{2 \cdot 2 \cdot 7}{3 \cdot 5}$$

Try this one. I've shown the Keep, Change, Flip.

b)
$$\frac{4x + 12}{6x - 18} \div \frac{3x + 9}{5x - 15} = \frac{4x + 12}{6x - 18} \cdot \frac{5x - 15}{3x + 9}$$

Try another. Remember the Keep, Change, Flip.

c)
$$\frac{14x - 28}{4x + 4} \div \frac{x^2 + 2x - 8}{2x + 2}$$

Look at this problem:

$$\frac{\frac{2}{3}}{\frac{4}{7}}$$

It is still just division:

$$\frac{2}{3} \div \frac{4}{7}$$

d) Finish it by using Keep, Change, Flip.

Try these. They are still just division. Keep the top. Change to multiplication and flip the bottom.

e) $\dfrac{\dfrac{6x-18}{x^2-4}}{\dfrac{3x^2-27}{x-2}}$

f) $\dfrac{\dfrac{x^2-4}{2x}}{\dfrac{x^2+x-6}{x^3+6x^2+9x}}$

Active Lesson: 8.3a

In this activity, we are going to take a look at how we can add and subtract rational expressions. Rational expressions are simply algebraic fractions and so it will follow all the standard rules for fractions.

Add the following:

a)
$$\frac{1}{5} + \frac{2}{5} =$$

If you have a common denominator, simply add the numerators and keep the denominator. Nothing changes with algebraic fractions. If the denominator is the same, just add the numerators. Try these:

b)
$$\frac{2}{x+5} + \frac{3}{x+5} =$$

c)
$$\frac{3y}{y-7} + \frac{4y}{y-7} =$$

For our next type of problem, let's return to something familiar. Add the following fractions and then reduce your answer.

d)
$$\frac{3}{8} + \frac{3}{8} =$$

Although you probably reduced in your head, what you are really doing is making the prime factorization for the numerator and the denominator and then cancelling any matching factors.

$$\frac{2 \cdot 3}{2 \cdot 2 \cdot 2} = \frac{3}{4}$$

We follow the same process when adding rational expressions. First add the fractions and then check to see if they can be reduced. Try these:

e) $\dfrac{x}{x^2 - 9} + \dfrac{3}{x^2 - 9} =$

f) $\dfrac{y - 7}{y^2 - 25} + \dfrac{2}{y^2 - 25} =$

g) $\dfrac{x^2 + x + 3}{x^2 - 81} + \dfrac{-7x - 30}{x^2 - 81} =$

The process for subtracting rational expressions is also the same as subtracting simple fractions. Subtract this fraction.

h)

$$\frac{5}{7} - \frac{4}{7} =$$

In the same way, subtract these rational expressions.

i) $\dfrac{5}{x-8} - \dfrac{3}{x-8}$

j) $\dfrac{y}{y+5} - \dfrac{2}{y+5}$

And after we subtract fractions, we check to see if they can be reduced. Subtract this fraction:

k)

$$\frac{9}{14} - \frac{3}{14} =$$

Now, do the same with rational expressions. First subtract them. Then factor and simplify.

l) $\dfrac{7y}{y^2+5y} - \dfrac{6y}{y^2+5y} =$

m) $\dfrac{2z-4}{z^2-16} - \dfrac{z}{z^2-16} =$

Finally, we need to subtract some more complicated rational expressions. Before we do, multiply the following:

n) $-1(3x + 5) =$

o) $-1(-x^2 + 5x - 8) =$

Nothing changes if we have only a negative sign and not the negative one. Multiply the following:

p) $-(3x + 5) =$

q) $-(-x^2 + 5x - 8) =$

So, when this becomes rational expressions, the subtraction sign acts like the negative in the problems above. We distribute it to everything in the numerator which follows:

$$\frac{2x + 4}{x - 4} - \frac{x + 6}{x - 4} = \frac{2x + 4 - x - 6}{x - 4} = \frac{x - 2}{x - 4}$$

Try some:

r) $\dfrac{3y - 8}{y + 5} - \dfrac{y - 10}{y + 5} =$

On this problem, once you've subtracted, check to see if you can reduce.

s) $\dfrac{4r - 8}{r^2 - 36} - \dfrac{3r - 2}{r^2 - 36} =$

Try one more. Subtract and then see if you can reduce. (It may or may not.)

t) $\dfrac{3x^2 + 5}{x^2 + 6x - 16} - \dfrac{2x^2 + 21}{x^2 + 6x - 16} =$

Algebra I

Active Lesson: 8.3b

a) We saw in an earlier activity, that we can turn a polynomial around by factoring. What number do we factor out of a binomial like $(2 - x)$ in order to turn it into $(x - 2)$?

Flip around the following binomials by factoring:

b) $(15 - y)$

c) $(2 - z)$

d) $(3 - x^2)$

We can use this trick to help us add and subtract rational expressions. Notice that these two fractions almost have the same denominator, however the one on the right is the wrong way around. So, I'm going to use the trick.

$$\frac{3}{x - 8} + \frac{7}{8 - x}$$

$$\frac{3}{x - 8} + \frac{7}{-1(x - 8)}$$

And while that is the right trick. With fractions, that -1 can just be turned into a negative sign and put out front.

$$\frac{3}{x - 8} - \frac{7}{x - 8} = \frac{-4}{x - 8} \text{ or } -\frac{4}{x - 8}$$

On these problems, do the same trick I did and then solve.

e) $\dfrac{6}{z - 7} + \dfrac{2}{7 - z}$

f) $\dfrac{3v}{v - 1} + \dfrac{v}{1 - v}$

429

When this problem is turned into subtraction, be sure to distribute it to both the $3x$ and the 2.

g) $\quad \dfrac{5x+1}{x-12} + \dfrac{3x+2}{12-x}$

Here, be sure to simplify after you add the fractions.

h) $\quad \dfrac{x}{x^2-9} + \dfrac{3}{9-x^2}$

Try one more. Remember to distribute the negative sign. Factor after you've combined.

i) $\quad \dfrac{2x^2+x-2}{x^2-4} + \dfrac{x^2+3x+6}{4-x^2}$

And if we start with a subtraction problem, the trick will turn it into an addition problem.

$$\dfrac{2x}{x-5} - \dfrac{5}{5-x}$$

This becomes:

$$\dfrac{2x}{x-5} + \dfrac{5}{x-5} = \dfrac{2x+5}{x-5}$$

Do the same trick on these problems.

j) $\dfrac{4}{x-5} - \dfrac{3}{5-x}$

k) $\dfrac{3y}{y-9} - \dfrac{4}{9-y}$

l) $\dfrac{x+2}{x^2-16} - \dfrac{2}{16-x^2}$

m) $\dfrac{y^2+30}{y^2-36} - \dfrac{12y+6}{36-y^2}$

To understand fractions, you need to remember that they are like slices of pie.

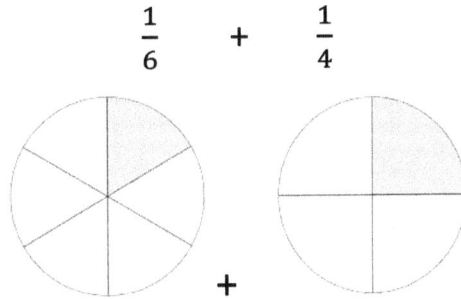

$$\frac{1}{6} \quad + \quad \frac{1}{4}$$

+

You can't simply add these because they are sliced differently. To fix the problem, we find something called a Least Common Denominator. To find it, factor the denominators.

$$\frac{1}{2\cdot 3} \quad + \quad \frac{1}{2\cdot 2}$$

Now, we just make the denominators match. The left needs one more 2 and the right needs one more 3. But if you give it to the bottom, you must also give it to the top.

$$\frac{1\cdot 2}{2\cdot 3\cdot 2} + \frac{1\cdot 3}{2\cdot 2\cdot 3} \;=\; \frac{2}{12} \quad + \quad \frac{3}{12} \;=\; \frac{5}{12}$$

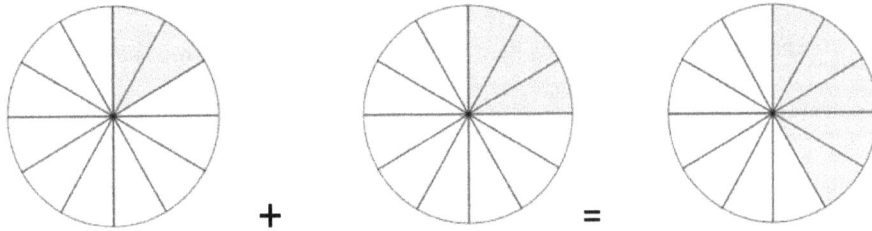

+ =

The idea works exactly the same in algebra.

$$\frac{1}{6a} \quad + \quad \frac{1}{4b}$$

$$\frac{1}{2\cdot 3\cdot a} \quad + \quad \frac{1}{2\cdot 2\cdot b}$$

Make the denominators match. (Remember, if you give something to the denominator, you must also give it to the numerator.) Then, once the denominators match, add the fractions like we did in the last section.

a) Finish the problem in the space below:

$$\frac{1}{2 \cdot 3 \cdot a} + \frac{1}{2 \cdot 2 \cdot b}$$

b) What is the Least Common Denominator? (Just multiply all of the, now matching, factors back together.)

Try this one. (The fraction on the left will need $2x$ multiplied to the top and bottom. The fraction on the right will need a y multiplied to the top and bottom.)

c) $\dfrac{1}{5x^3y^2} + \dfrac{1}{10x^4y}$

The process is the same even if the fractions have polynomials on the bottom. Factor everything first.

$$\frac{x}{x^2 - 9} + \frac{2}{3x + 9} = \frac{x}{(x-3)(x+3)} + \frac{2}{3(x+3)}$$

The fraction on the left needs a 3. The fraction on the right needs $(x - 3)$. In the space below, make the denominators match. (Don't forget to give to the tops!)

d)
$$\frac{x}{(x-3)(x+3)} + \frac{2}{3(x+3)}$$

e) What is the LCD? (When the denominators are complicated like this, it is easier, and better, not to multiply them back together.)

The only way to add the fractions is to clean up the numerators. I've multiplied them out below. Finish adding the fractions.

f)

$$\frac{3x}{3(x-3)(x+3)} + \frac{2x-6}{3(x-3)(x+3)}$$

Try one on your own. After you get the common denominator, multiply out the numerator, and then add the fractions.

g)

$$\frac{x}{x^2+7x+12} + \frac{3}{2x+8}$$

h) What is the LCD? (Don't multiply the denominator back together. Leave it in factored form.)

On this problem, I've done the factoring in the denominator for you. After you add the fractions, you need to factor the new numerator. It will simplify.

i)

$$\frac{x}{(x+3)(x+1)} + \frac{2}{(x+5)(x+1)}$$

Here's one final problem. Try the entire thing on your own.

j) $$\frac{4}{x^2 + 2x - 3} + \frac{x}{x^2 - 3x + 2}$$

Algebra I

Active Lesson: 8.4b

Subtract the following polynomials: (Remember that the negative sign before the second polynomial must be distributed to every term that follows.)

a) $(6x + 15) - (3x + 12)$

b) $(x^2 + 3x - 2) - (5x + 7)$

c) $(15x - 2) - (3x - 11)$

We are going to continue to work with rational expressions. Previously, we added. Now, we will subtract. As before, our fractions need a common denominator. Only one thing changes, we must remember to distribute the negative sign just as we do when we subtract polynomials.

$$\frac{5}{(x-2)} - \frac{3}{(x+3)}$$

The left needs the $(x + 3)$ and the right needs the $(x - 2)$. If we give them to the bottom, we must give them to the top.

$$\frac{5(x+3)}{(x-2)(x+3)} - \frac{3(x-2)}{(x-2)(x+3)}$$

To subtract these rational expressions, we first need to multiply out the tops (numerators).

$$\frac{5x+15}{(x-2)(x+3)} - \frac{3x-6}{(x-2)(x+3)}$$

The denominators are a match, so we can combine the numerators. But the subtraction sign will change the sign of everything which follows.

$$\frac{5x + 15 - 3x + 6}{(x-2)(x+3)}$$

$$\frac{2x + 21}{(x-2)(x+3)}$$

We would check to see if the numerator factors, but it doesn't, and so we are done.

Try one on your own.

d) $\dfrac{7}{x+5} - \dfrac{4}{x-2}$

As we saw before, we will often need to factor the denominators first. I've done the first step for you.

$$\frac{x}{x^2-9} - \frac{2}{3x+9} = \frac{x}{(x-3)(x+3)} - \frac{2}{3(x+3)}$$

Finish this problem. The left side needs a 3 added to the top and bottom. The right side needs an
e) $(x-3)$ added.

Try this problem. You will have three binomials in the denominator. At the end, you will be able to reduce.

f) $$\frac{4}{(x-2)(x+3)} - \frac{3}{(x-2)(x+2)}$$

Work one final problem. Factor the denominators first. At the end of the problem, you will be able to reduce.

g) $$\frac{x}{x^2-16} - \frac{1}{2x-8}$$

438

Algebra I

Active Lesson: 8.5a

If we start with this problem:

$$\frac{3}{4} - \frac{1}{6} + 2$$

The first thing we must do is get a common denominator. In order to do that, we must remember that we should turn the 2 into a fraction. I've don't that part for you. Finish the rest.

a)

$$\frac{3}{4} - \frac{1}{6} + \frac{2}{1}$$

Now, simplify this:

b)

$$\frac{1}{3} + \frac{1}{2}$$

Next, suppose I started with something as complicated as this. (It is called a complex fraction.)

$$\frac{\frac{3}{4} - \frac{1}{6} + 2}{\frac{1}{3} + \frac{1}{2}}$$

It looks terrible. However, you've already done the hard part. You simplified the numerator, and you simplified the denominator. So, you have a fraction divided by a fraction. Therefore, Keep Change Flip is the final step. In the space below, do the work to finish the problem.

c)

If this is an algebra problem, it looks even worse:

$$\frac{1 - \frac{1}{x}}{\frac{x}{y} + \frac{y}{x}}$$

But we know how to do everything here, we just need to break it into steps.

Solve this problem:

d)

$$1 - \frac{1}{x}$$

Now, solve this problem:

e)

$$\frac{x}{y} + \frac{y}{x}$$

Take your two answers and put them over each other.

f)

$$\frac{Answer \#1}{Answer \#2} = \frac{\square}{\square}$$

Then, Keep, Change, Flip and simplify. Do so in the space below.

g)

Try one more. Simplify:

h)

$$\frac{\frac{1}{x} + \frac{1}{y}}{x - \frac{y^2}{x}}$$

Algebra I

Active Lesson: 8.5b

In this activity, we are going to see a second approach to simplifying complex rational expressions. I do not teach this approach in a classroom because I prefer the previous method. However, if you feel confident, this approach can be faster.

Here is the problem we began with in the previous activity:

$$\frac{\frac{3}{4}-\frac{1}{6}+\frac{2}{1}}{\frac{1}{3}+\frac{1}{2}}$$

Instead of breaking it into three problems (top, bottom, and then division), we are going to find a Least Common Denominator for every fraction in the problem. (If you aren't sure how, look back on our previous activity regarding LCD's.) The least number that all the denominators could become is 12. Next, we will multiply both the top (numerator) and the bottom (denominator) by 12.

$$\frac{12\left(\frac{3}{4}-\frac{1}{6}+\frac{2}{1}\right)}{12\left(\frac{1}{3}+\frac{1}{2}\right)}=\frac{12\cdot\frac{3}{4}-12\cdot\frac{1}{6}+12\cdot\frac{2}{1}}{12\cdot\frac{1}{3}+12\cdot\frac{1}{2}}$$

It reduces down to:

$$\frac{9-2+24}{4+6}=\frac{31}{10}$$

As you can see, that was much faster. Here we will use the problem in the context of algebra:

$$\frac{1-\frac{1}{x}}{\frac{x}{y}+\frac{y}{x}}$$

The LCD of all the fractions would be xy. Multiply xy times the top and bottom.

a)

$$\frac{xy\left(1-\frac{1}{x}\right)}{xy\left(\frac{x}{y}+\frac{y}{x}\right)}$$

441

Try another. The LCD for the fractions would again be xy:

b) $$\dfrac{\dfrac{1}{x}+\dfrac{1}{y}}{x-\dfrac{y^2}{x}}$$

Let's look at a problem that involves binomials.

$$\dfrac{\dfrac{5}{x-3}}{\dfrac{7}{x+3}-\dfrac{4}{x^2-9}}$$

To find an LCD for all the fractions, we need to factor all the denominators. The only one which needs work is x^2-9. We get:

$$\dfrac{\dfrac{5}{x-3}}{\dfrac{7}{x+3}-\dfrac{4}{(x-3)(x+3)}}$$

The LCD is: $(x-3)(x+3)$. Now, we multiply by the LCD on the top and bottom.

$$\dfrac{(x-3)(x+3)\left(\dfrac{5}{x-3}\right)}{(x-3)(x+3)\left(\dfrac{7}{x+3}-\dfrac{4}{(x-3)(x+3)}\right)}$$

This isn't as hard as it looks, but this is why I don't use this method. After multiplying, we would get:

$$\dfrac{(x+3)(5)}{(x-3)(7)-4}$$

c) Finish simplifying. (The 5 and 7 aren't on the typical side, but they still need to be distributed.)

Try this problem on your own.

d)

$$\dfrac{\dfrac{6}{(x+5)}}{\dfrac{8}{(x-5)}-\dfrac{3}{x^2-25}}$$

I've created some fractions using pie slices. Beside them, write the real math equation. (You don't need to solve it yet.)

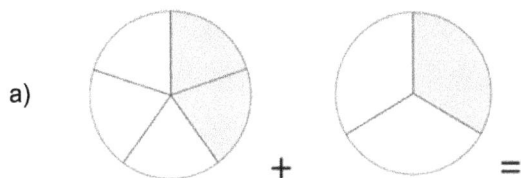

a)

$+$ $=$

The same two pies have now been cut so that they are sliced the same, i.e. a common denominator. Beside them, write the real math problem.

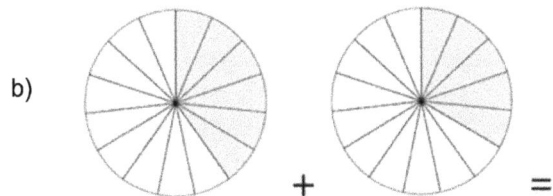

b)

$+$ $=$

Obviously, the answer to the problem is $\frac{11}{15}$. But suppose it had started as an algebra problem.

$$\frac{x}{5} + \frac{1}{3} = \frac{11}{15}$$

First, I'm going to get a common denominator.

$$\frac{x \cdot 3}{5 \cdot 3} + \frac{1 \cdot 5}{3 \cdot 5} = \frac{11}{15}$$

$$\frac{3x}{15} + \frac{5}{15} = \frac{11}{15}$$

Now, I solve:

$$3x + 5 = 11$$

$$3x = 6$$

$$x = 2$$

I want you to notice something very important. Once you have gotten the slices the same size (a common denominator), the slices have nothing to do with it.

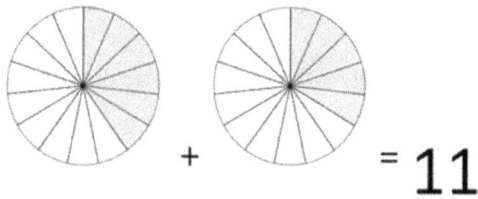

 + = **11**

All I needed to solve was:

$$3x + 5 = 11$$

I didn't do anything with the 15 in the denominator. And this is always true. Get a common denominator and then just solve the equation on the top. Look at this one:

$$\frac{x}{4} + \frac{1}{3} = 2$$

So, I get a common denominator.

$$\frac{x \cdot 3}{4 \cdot 3} + \frac{1 \cdot 4}{3 \cdot 4} = \frac{2 \cdot 4 \cdot 3}{1 \cdot 4 \cdot 3}$$

$$\frac{3x}{12} + \frac{4}{12} = \frac{24}{12}$$

Now solve this. Remember, the 12's have nothing to do with finding the value of x.

c)

Try this one on your own. Solve this equation for x.

d)
$$\frac{x}{2} + \frac{1}{4} = \frac{6}{3}$$

More difficult versions of the problem have variables on the bottom.

$$1 - \frac{3}{x} = \frac{4}{x^2}$$

First, make the 1 a fraction:

$$\frac{1}{1} - \frac{3}{x} = \frac{4}{x^2}$$

Next, get them all over a denominator of x^2. After you have a common denominator, you will need to solve the problem using the zero-product property (one side of the equation equals zero). Show your

e)   work in the space below:

On this next problem, first factor the denominator on the left. Then, get each of the denominators to match. If you give to the bottom, don't forget to give to the top. Once the denominators match, you don't need them anymore; just solve the numerator.

f) $$\frac{x+5}{x^2 + 2x - 15} = \frac{7}{x+5} - \frac{5}{x-3}$$

Some time back, we saw that any value which would cause the denominator to be zero is not allowed. What values of x are not permitted to be answers to this problem?

$$\frac{x-17}{x^2+x-6}=\frac{4(x-2)}{x+3}-\frac{2(x+3)}{x-2}$$

Now, get a common denominator and solve. You will get an answer that is not allowed. Therefore, your answer would be no solution.

g)

Any time you solve rational equations, you must check to see if any of your answers are not permitted. If they aren't allowed, ignore them as a possible solution.

Solve the following equations for the variable indicated:

a)   Solve for W.

$$A = LW$$

b)   Solve for M.

$$F = MA$$

In this activity, we want to take a rational expression and solve for a variable. For instance, if we start with an equation like this:

$$\frac{D}{R} = T$$

And we wanted to solve for R, we first need to get R out of the denominator. To do this, we multiply both sides by R.

$$R \cdot \frac{D}{R} = T \cdot R$$

And we wind up with:

$$D = TR$$

Now, to solve for R, we simply divide the T away from both sides.

$$\frac{D}{T} = \frac{TR}{T}$$

Or:

$$\frac{D}{T} = R$$

Try these:

Solve for W.

c) $\dfrac{A}{W} = L$

Solve for M.

d) $\dfrac{F}{M} = A$

This next problem is going to show a valuable trick that will carry over into advanced levels of algebra.

Solve for y.

$$x = \frac{y}{y+2}$$

The difficulty here is that y is in two places. We will never be able to get y alone unless we can figure out how to combine them. Here's the trick. The denominator $y + 2$ is all one number:

$$x = \frac{y}{(y+2)}$$

So, we can multiply both sides by that number.

$$(y+2)x = \frac{y}{(y+2)}(y+2)$$

And, therefore:

$$(y+2)x = y$$

This is simpler, but we still need to combine the $y's$. We can do this by first distributing the x and then getting all the y terms alone on one side.

$$yx + 2x = y$$

Now, I just move the 2x to the right and the y to the left.

$$yx - y = -2x$$

Next, the left side has a greatest common factor of a y. We can pull that out:

$$y(x - 1) = -2x$$

Finally, $(x - 1)$ is a number and we can divide the number across to the other side.

$$y = \frac{-2x}{(x - 1)}$$

Try this problem.

Solve for x.

e) $\quad y = \dfrac{x}{x - 3}$

The next type of problem will follow the ideas for solving rational equations. If we get a common denominator, the denominator won't matter.

Solve for y.

$$\frac{1}{x} + \frac{1}{y} = 5$$

$$\frac{1y}{xy} + \frac{1x}{xy} = \frac{5xy}{xy}$$

Now, we can disregard the denominator.

$$y + x = 5xy$$

To solve for y we need the trick from the last example. We will get all the y's on the same side.

$$y - 5xy = -x$$

Pull out the GCF of y. (Sometimes there will be a GCF with more than just the variable we want. However, for the trick to work properly, we just take out the variable we want to get alone.)

$$y(1 - 5x) = -x$$

And, as before, $(1 - 5x)$ is a number, so we can divide it down to the other side.

$$y = \frac{-x}{(1 - 5x)}$$

Try this problem:

Solve for y. Get a common denominator of $5xy$. Then, you can disregard the denominator.

f) $\dfrac{1}{x} + \dfrac{1}{5} = \dfrac{1}{y}$

Algebra I

Active Lesson: 8.7a

Earlier in the course, I reminded you of a strategy for solving proportions like this one.

$$\frac{3}{5} = \frac{x}{100}$$

The trick was to lasso, like this:

Then you had:

$$3 \cdot 100 = 5 \cdot x$$

$$\text{Or } 300 = 5x$$

$$x = 60$$

Work these two.

a) $\dfrac{2}{12} = \dfrac{x}{84}$

b) $\dfrac{x}{11} = \dfrac{90}{165}$

Radical expressions can be proportions too. If there is one ratio (fraction) on the left side of the equation and one on the right, it is the same as a proportion. Let's see why. I've started one below, working it like we did in the last section, by getting a common denominator.

$$\frac{x}{x + 14} = \frac{5}{7}$$

$$\frac{x \cdot 7}{(x + 14) \cdot 7} = \frac{5 \cdot (x + 14)}{7(x + 14)}$$

c) Now, finish this. (Remember, once you have a common denominator, we only need to solve the numerator.)

Now, look what happens when we lasso.

$$7 \cdot x = 5 \cdot (x + 14)$$

It is exactly the same problem we already solved. Solve the following radical expressions by using this lasso method.

d) $\dfrac{x}{x+3} = \dfrac{3}{4}$

e) $\dfrac{x-2}{3} = \dfrac{x+5}{4}$

f) $\dfrac{2y-4}{8} = \dfrac{3y+2}{4}$

Algebra I

Active Lesson: 8.7b

Solve the following proportion:

a)

$$\frac{20}{5} = \frac{110}{x}$$

In this section, we will use proportions to solve word problems.

A pediatrician gives 5ml of medicine for every 20lbs that a child weights. If a child weighs 110lbs how many ml of medicine should the child receive?

The key is to keep the ratio the same.

$$\frac{lbs}{ml} = \frac{lbs}{ml}$$

You could choose to flip both ratios, and that would work too. But you must remain consistent. In this problem 5 goes with 20 and 110 goes with x. So:

$$\frac{20}{5} = \frac{110}{x}$$

Notice that this is the proportion that you already solved.

Try this one on your own.

b) A pediatrician gives 6ml of medicine for every 30lbs that a child weights. If a child weighs 120lbs how many ml of medicine should the child receive?

The idea really doesn't change no matter the context. Just be sure to be consistent with your ratios. Try these problems.

c) An energy drink comes in two sizes. The 12 ounce size has 300 calories. How many calories are in the 16 ounce size?

d) Currently, one U.S. dollar is the same as 1.40 Canadian dollars. If you want to exchange $450 U.S. dollars, how many Canadian dollars would you receive?

The triangles below are in a special relationship.

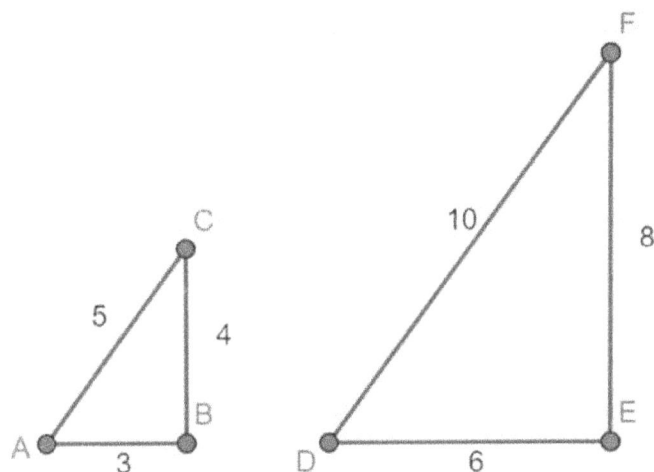

a)   Look at their matching sides. What is special about their relationship?

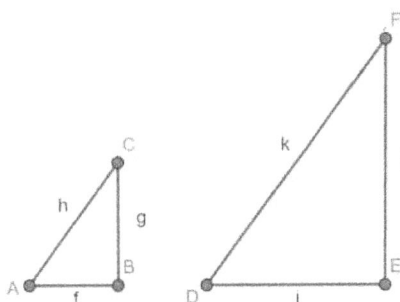

When the sides of two shapes are in a similar ratio, we call the triangles similar triangles. Practically speaking, what this means is that we can set up a proportion:

$$\frac{i}{f} = \frac{j}{g} = \frac{k}{h}$$

It doesn't matter which triangle goes on top and which goes on bottom, just stay consistent. (I put all the sides from the larger triangle on top, but you could have flipped it.)

We will use this fact, to solve problems in which we know three of the four sides. Because each of the ratios are the same, we can use the formula in any arrangement.

$$\frac{i}{f} = \frac{j}{g} \qquad\qquad \frac{j}{g} = \frac{k}{h} \qquad\qquad \frac{i}{f} = \frac{k}{h}$$

Here's a problem.

$\triangle ABC$ is similar to $\triangle DEF$. Find the missing side.

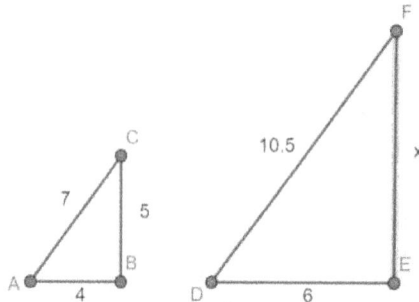

Because these are similar triangles, we can set up a proportion to find x. As long as we have three of four sides, and we keep them in their consistent ratio, we can make this equation:

$$\frac{4}{6} = \frac{5}{x}$$

Again, we need to stay consistent. I've done the smaller sides on top. We also could have made the equation:

$$\frac{7}{10.5} = \frac{5}{x}$$

b) Use the lasso method to solve this equation and find the missing side:

$$\frac{4}{6} = \frac{5}{x}$$

Try one on your own.

c) $\triangle ABC$ is similar to $\triangle DEF$. Find the missing side.

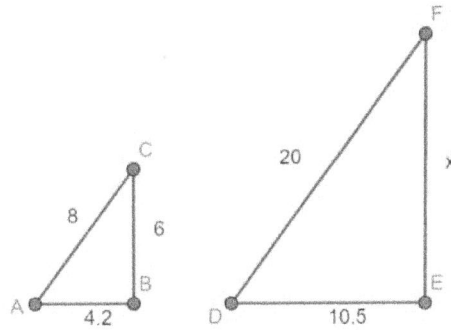

There are many applications of the concept of similar triangles. Another is when working with the scale on a map. The triangle on the right is the actual distance between three towns. The triangle on the left

d) is from a scaled map. Set up a proportion and find the missing distance between Rush and Holt.

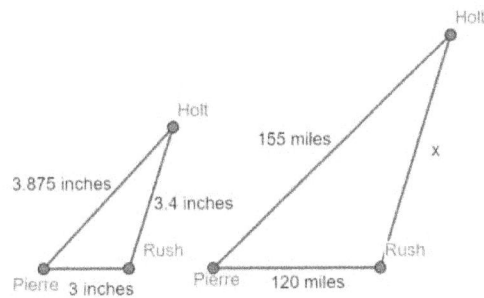

When two triangles are built "on top" of each other, sharing the same vertex, they are also similar triangles.

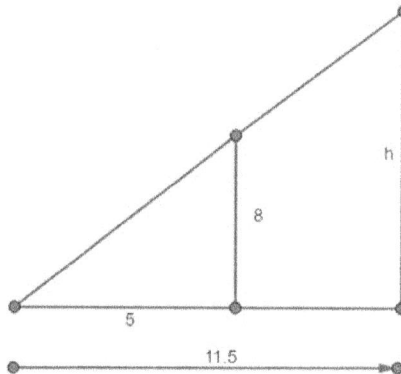

Here, we can set up a proportion to solve for the missing height of the larger triangle. This is the proportion which we can create:

$$\frac{5}{11.5} = \frac{8}{h}$$

e) Solve for h.

A common application of this problem would be a word problem about shadows.

f)

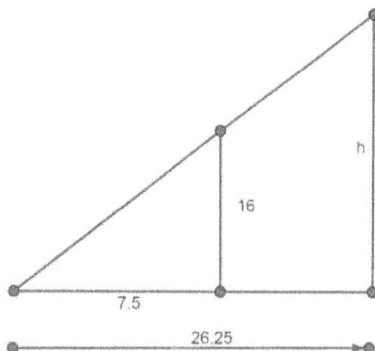

The tops of a flagpole and a building are each giving a shadow. The flagpole is 16 feet tall and is giving a 7.5 foot shadow. The building is giving a 26.25 foot shadow. How tall is the building?

Algebra I

Active Lesson: 8.8a

In this section, we are going to revisit a type of word problem we saw in earlier material. This time, we will find that the word problems are rational equations. The formula we will need is:

$$D = r \cdot t$$

a) Solve this equation for time.

Let's look at a problem.

A small airplane travelled 300 miles against the wind in the same amount of time as it could travel 400 miles with the wind. If the rate of the wind was 20 mph, what was the rate of the plane?

Previously, we saw how the wind can help speed up the plane, $x + 20$, or it can slow down the plane, $x - 20$. Here is what out dictionary looks like so far:

Math	English
x	Rate of the plane

	Rate	Time	Distance
With the Wind	x+20		400
Against the Wind	x-20		300

The first time we worked these problems, we would use the rate and the time column to create the distance column. Here, we must use the rate and distance column to create the time column. When you solve $D = r \cdot t$ for time, you get time equals distance divided by rate. So, we can fill in the time column in our dictionary.

	Rate	Time	Distance
With the wind	x+20	$\dfrac{400}{x+20}$	400
Against the wind	x-20	$\dfrac{300}{x-20}$	300

461

To make an equation, we know that the time the plane travelled was the same in both directions. So, to have the left and right sides tell the same story, we have:

$$\frac{400}{x + 20} = \frac{300}{x - 20}$$

b) Use the lasso method to solve this proportion for x, the rate of the plane.

Try one on your own.

c) A small airplane travelled 500 miles against the wind in the same amount of time as it could travel 600 miles with the wind. If the rate of the wind was 10 mph, what was the rate of the plane?

Let's look at another variation.

An athlete spent four hours training for a biathlon. She swam for 6 miles and biked for 10 miles. If she bikes 8 mph faster than she swims, find the rates at which she bikes and swims.

Here is the start of our dictionary:

	Rate	Time	Distance
Biking	x+8		10
Swimming	x		6

As before, we create the time column by dividing the distance by the rate and get:

	Rate	Time	Distance
Biking	x+8	$\frac{10}{x+8}$	10
Swimming	x	$\frac{6}{x}$	6

To make the left and right side of the equation tell the same story, we know the entire time she spent training was 4 hours. So, if we add up the time biking and swimming, we can set it equal to the 4 hours training.

$$\frac{10}{x+8} + \frac{6}{x} = 4$$

d) Solve this equation. Get the denominators to match, including under the 4. If you give something to the bottom, you must give it to the top. Then, you can ignore the bottom. You will then need to get one side equal to zero and factor to solve.

e) You should have gotten two values for x. Disregard the answer of -6 because a rate can't be negative. What is the rate of biking and what is the rate of swimming?

This final problem is very difficult. I wouldn't typically assign a problem like this in class, but I will show you how it can be done.

A hiker travels uphill 10 miles and downhill 6 miles. His rate is 3 mph slower going uphill. Traveling uphill also takes 1 hours longer.

Here's our dictionary so far:

	Rate	Time	Distance
Uphill	x-3		10
Downhill	x		6

As before, we find the time column by dividing:

	Rate	Time	Distance
Uphill	x-3	$\frac{10}{x-3}$	10
Downhill	x	$\frac{6}{x}$	6

Setting up an equation here is more difficult than normal. We don't have a total time, so we can't add the time column. And, the time for uphill and downhill are not equal. This problem is so difficult because all we know is that the uphill time took 1 hours longer. To have the right and left sides of the equation telling the same story, we would need to subtract 1 from the uphill time.

$$\frac{6}{x} = \frac{10}{x-3} - 1$$

This is now two times which are equal. To solve, get a common denominator, including for the 1. Try it. You should get $x = 9 \ or \ x = -2$. Since we can't have a negative rate, only the $x = 9$ value will be true.

Algebra I

Active Lesson: 8.8b

We want to solve the following proportion:

A teenager can rake a large yard in 4 hours. How much of the yard could she rake in 1 hour?

Here is our set up:

$$\frac{1 \; job}{4 \; hours} = \frac{x \; job}{1 \; hour}$$

a) Solve this for x using the lasso approach.

b) The teenager's little brother could rake a large yard in 12 hours. How much of the yard could he rake in 1 hour? Set up the proportion and solve.

In this section, we are looking at another type of word problem. This time it involves the rate at which work can be done. Here's what a problem will look like.

A teenager can rake a large yard in 4 hours. Her little brother can rake a large yard in 12 hours. How long would it take them to rake a large yard if they worked together?

Here's the dictionary that we want to complete:

Math	English
	Hourly rate of the teenager
	Hourly rate of the little brother
	Hourly rate together

If you look at the proportions that we completed at the start of this activity, we see that an hourly rate is simply the time it would take them to finish the work alone underneath a 1.

So, our table looks like this:

Math	English
$\frac{1}{4}$	Hourly rate of the teenager
$\frac{1}{12}$	Hourly rate of the little brother
	Hourly rate together

And if they work together, there is some hourly rate, which we don't know. So:

Math	English
t	Total time to complete the job together

We add it as a rate to our dictionary:

Math	English
$\frac{1}{4}$	Hourly rate of the teenager
$\frac{1}{12}$	Hourly rate of the little brother
$\frac{1}{t}$	Hourly rate together

Next, we need an equation. If they work together, we can add their rates. And that must equal their "together" rate.

$$\frac{1}{4} + \frac{1}{12} = \frac{1}{t}$$

To find the total time it would take them to rake the yard together, we need to solve for t. We know how
c)   to do this. Get a common denominator among each fraction and then the denominator doesn't matter.

Try another problem on your own.

d) It takes an employee four hours to complete a project. A second employee can complete the project in six hours. How long would it take for them to complete the project if they work together?

(You will get a decimal answer.)

We can work a similar type of problem where we know the rate of them working together.

A new employee is working together with their trainer to finish a project. Together they can complete the project in 2 hours. Alone the trainer can finish the project in 3 hours. How long would it take the new employee to finish the job if they were working alone?

Here's our dictionary:

Math	English
$\frac{1}{t}$	Rate of New Employee
$\frac{1}{3}$	Rate of Trainer
$\frac{1}{2}$	Rate Together

So, our equation becomes:

$$\frac{1}{t} + \frac{1}{3} = \frac{1}{2}$$

e) Solve for t and give the time it would take the new employee to finish the job alone.

Work this problem on your own.

f)   A new employee is working together with their trainer to finish a project. Together they can complete the project in 3 hours. Alone the trainer can finish the project in 4 hours. How long would it take the new employee to finish the job if they were working alone?

Solve each of the following equations for k.

a) $5 = k(10.5)$

b) $8 = k(16.8)$

c) $12 = k(25.2)$

d)   What did each of the previous values of k have in common?

When two variables have a constant ratio, they are said to be in direct variation.

$$y = kx$$

Here, I've solved for k:

$$\frac{y}{x} = k$$

For every pair of values of x and y, the ratio would always be the same. We call that ratio, k, the constant of variation. Practically speaking, if the ratio is always the same, it means that if one variable goes up then the other variable must also go up. Here is what a problem would look like:

If y varies directly with x when $y = 15$ and $x = 2.5$, find the equation which relates x and y.

Because they told us that direct variation exists then our formula must be based on $y = kx$. But to create the formula, we must have the value of k. Since we have values for x and y, we can sub in and find k.

$$15 = k(2.5)$$

$$k = 6$$

If that is the constant of variation then it is true for all pairs of x and y, which gives us the equation we need:

$$y = 6x$$

Try a problem on your own.

e)　If y varies directly with x when $y = 8$ and $x = 1.6$, find the equation which relates x and y.

Word problems involving direct variation are straightforward.

The calories burned, c, is in direct variation with the time exercising, t. If you burned 500 calories in 80 minutes of exercising, find the equation of direct variation.

To find the equation, we need to first find the value of k.

$500 = k(80)$

f)　Find k and show the equation of direct variation.

The questions will often have a second part.

g)　If the next day, you exercised for 95 minutes, how many calories did you burn.

Insert 95 into your direct variation equation to solve for c.

Try this problem.

h)   The miles traveled in your car, m, varies directly with the time spent driving, t. If you travelled 600 miles in 15 hours, find the following:

    a)   The equation for direct variation.
    b)   The number of miles you would have driven in 27 hours.

Finally, direct variation can occur between a variable and the square of another. Here's an example.

i)   The area of a circle, A, varies directly with the square of the radius, r^2. A circle with an area of $12.56\ m^2$ has a radius of $2\ m$.

    a)   Give the equation for direct variation.

This time our direct variation relationship is $A = kr^2$. Use the information given to find k and then write out the equation for direct variation.

    b)   If the radius of circle is 6 inches, find the area.

Solve each of the following equations for k.

a) $6 = \dfrac{k}{2}$

b) $4 = \dfrac{k}{3}$

c) $12 = \dfrac{k}{1}$

d) What did each of the previous values of k have in common?

Solve this equation for k.

e)

$$y = \frac{k}{x}$$

When two variables had a constant ratio, we said that they vary directly. When two variables have a constant product, we say that they vary indirectly. To maintain the constant product, practically speaking, it means that if the value of one variable goes up then the other must go down, and vice versa.

For every pair of values of x and y, the product would always be the same. We call that product, k, the constant of variation. Here is what a problem would look like:

If y varies indirectly with x when $y = 5$ and $x = 2.5$, find the equation which relates x and y.

Because they told us that indirect variation exists then our formula must be based on $y = \frac{k}{x}$. But to create the formula, we must have the value of k. Since we have a pair of values of x and y, we can sub in and find k.

$$5 = \frac{k}{2.5}$$

If that is the constant of variation then it is true for all pairs of x and y, which gives us the equation we need:

$$y = \frac{12.5}{x}$$

Try a problem on your own.

f)   If y varies indirectly with x when $y = 3$ and $x = 2.2$, find the equation which relates x and y.

As we saw with direct variation, word problems involving indirect variation are straightforward.

Your car's value, v, varies indirectly with age, a. You bought a five-year-old car for $10,000.

     Write the equation of variation.

Just as before, we plug into our inverse formula to find the value of k.

$$v = \frac{k}{a}$$

$$10000 = \frac{k}{5}$$

$$50000 = k$$

g)   Now that you know the value of k, write the equation of variation.

h)   Use your equation of variation to find the value of your car after 10 years.

Here, plug 10 in for the value of the variable a and solve for v. (Remember, you now know the value of k.)

Try these problems on your own.

The driving time to get to your destination, t, varies indirectly with your car's speed, s. If it took you 3.5 hours to get to your destination when your speed was 60 miles per hour, find the following:

i)   The equation of variation.

j)   What would your driving time be if your speed was 75 miles per hour.

The time to complete a project at work, t, varies indirectly with the number of people working on the project, p. If it took a team of 6 people 14 days to complete the project, find the following:

k)   The equation of variation.

l)   If a team had 10 people, how long would it take to complete the project.

475

A square root is a shortcut for asking, "What number multiplies by itself to get this number?"

$$\sqrt{9} = \sqrt{3 \cdot 3} = 3$$

Give me the square root of the following:

a) $\sqrt{81}$ b) $\sqrt{121}$ c) $\sqrt{36}$ d) $\sqrt{49}$

Square roots are fairly easy, but what about cube roots ($\sqrt[3]{}$) or 4th roots ($\sqrt[4]{}$). Those aren't quite as simple. So, there is a trick. The first thing you need to do is make a prime factorization. I've done one for you.

$$27 = 3 \cdot 3 \cdot 3$$

Now, you try these. Make the prime factorization.

e) $8 =$

f) $125 =$

g) $27 =$

h) $16 =$

Suppose I had $\sqrt[3]{27}$. I would put the prime factorization inside the house.

$$\sqrt[3]{27} = \sqrt[3]{3 \cdot 3 \cdot 3}$$

Now, this is a <u>cube</u> root, so it takes <u>three</u> of the same prime factor to bring one out. Here's what I mean.

$$\sqrt[3]{\overbrace{3 \cdot 3 \cdot 3}}$$

I have one set of three of a kind, so I can bring a single 3 outside the house.

$$\sqrt[3]{27} = \sqrt[3]{3 \cdot 3 \cdot 3} = 3$$

Let me show you one more.

$$\sqrt[4]{81} = \sqrt[4]{3 \cdot 3 \cdot 3 \cdot 3}$$

$$\sqrt[4]{\overbrace{3 \cdot 3 \cdot 3 \cdot 3}}$$

I am doing <u>4</u>th roots, so I need <u>four</u> of a kind. Because I have 4 threes, I can bring one out.

$$\sqrt[4]{81} = \sqrt[4]{3 \cdot 3 \cdot 3 \cdot 3} = 3$$

Try these:

i) $\sqrt[3]{8} =$

j) $\sqrt[3]{125} =$

k) $\sqrt[4]{16} =$

l) $\sqrt[5]{32} =$

If you don't know a square root, the same trick can be used.

$$\sqrt{625} = \sqrt{5 \cdot 5 \cdot 5 \cdot 5}$$

A square root requires pairs to bring something out.

$$\sqrt{625} = \sqrt{5 \cdot 5 \cdot 5 \cdot 5}$$

But here there are two pairs, so I can bring out two 5's. When you bring them out you multiply those two fives together.

$$\sqrt{625} = \sqrt{5 \cdot 5 \cdot 5 \cdot 5} = 5 \cdot 5 = 25$$

Try a few using this method.

m) $\sqrt{256} =$

n) $\sqrt{196} =$

o)   Finally, look at this square root: $\sqrt{-9}$. We need two of the same number to multiply to get -9. Which of the following must be the answer:

   a) 3
   b) -3
   c) No Solution

Algebra I

Active Lesson: 9.1b

In our last activity, we saw how to turn numbers under a root into their prime factorization. Then, we made matches to determine how many pairs (for square roots) we are allowed to bring out from under the root. (In class, I often call the root the "house.") Here, we are going to extend the idea to problems which involve variables. The concept is the same. Below, I've turned variables into their prime factors.

$$\sqrt{x^2} = \sqrt{x \cdot x}$$

Then, since we are under a square root (which I call the world of two), we need pairs to bring one out.

$$\sqrt{x^2} = \sqrt{x \cdot x} = x$$

Try some:

a) $\sqrt{y^2} =$

(Here it extends to two variables, but the idea is the same.)

b) $\sqrt{r^2 s^2} =$

Higher variables follow the concept too. When you bring out multiple variables, multiply them back together.

$$\sqrt{x^4} = \sqrt{x \cdot x \cdot x \cdot x} = x \cdot x = x^2$$

Try some:

c) $\sqrt{y^6} =$

d) $\sqrt{x^8} =$

e) $\sqrt{m^4 n^6} =$

f) $\sqrt[3]{x^3} =$

It isn't required to write out all the variables and then circle the matches although it helps too. If you can see in your mind how they would pair up, that is also fine.

g) $\sqrt[4]{x^4 y^8} =$

h) $\sqrt{a^4 b^2 c^6} =$

We then extend the idea one more time by adding numbers. Nothing changes, simply do the root of the numbers and work the variables as we've done above.

$$\sqrt{9x^2} = \sqrt{3 \cdot 3 \cdot x \cdot x} = 3x$$

Try some:

i) $\sqrt{81y^2} =$

j) $\sqrt{25x^2 y^4} =$

k) $\sqrt{36m^4n^8} =$

And remember, if you don't know the square root of a number, you can always do the prime factorization trick.

l) $\sqrt{144a^{10}b^6} =$

m) $\sqrt{225x^{12}} =$

n) $\sqrt[3]{27m^3} =$

o) $\sqrt[3]{64x^3y^6} =$

Provide the prime factorization for each of the following:

a) $8 =$

b) $81 =$

c) $32 =$

When we work with roots, knowing the prime factorization is important because sometimes factors are left behind inside the "house." Here's what I mean.

$$\sqrt{27} = \sqrt{3 \cdot 3 \cdot 3}$$

Remember, it is a square root, so we circle a pair of threes.

$$\sqrt{3\,\boxed{(3 \cdot 3)}}$$

But when we bring them out, one three didn't partner up and so it is left behind.

$$\sqrt{3 \cdot 3 \cdot 3} = 3\sqrt{3}$$

Now you try:

d) $\sqrt{8} =$

e) $\sqrt[3]{81} =$

f) $\sqrt[4]{32} =$

The idea is the same with variables. Some people can simply see how many variables would partner up and come out, but if you can't, you can make the prime factorization. I've done that for you. Partner them up. Then, bring any out that can come out. I've finished the first one for you.

$$\sqrt{x^3} = \sqrt{x \cdot x \cdot x} = x\sqrt{x}$$

g) $\sqrt[3]{y^4} = \sqrt[3]{y \cdot y \cdot y \cdot y} =$

h) $\sqrt{x^5} = \sqrt{x \cdot x \cdot x \cdot x \cdot x} =$

i) $\sqrt[3]{y^7} = \sqrt[3]{y \cdot y \cdot y \cdot y \cdot y \cdot y \cdot y} =$

Try some on your own.

j) $\sqrt{x^7} =$

k) $\sqrt{8x^9} =$

l) $\sqrt{100y^5} =$

m) $\sqrt{180a^4} =$

The concept is the same with other roots. (Cube roots require three of a kind. Fourth roots require four of a kind.)

n) $\sqrt[3]{250y^4} =$

o) $\sqrt[4]{64a^4b^5} =$

Look at the following:

$$3 + x + y$$

p) Why can't the terms be combined?

Roots are like variables. And, for the same reason, these can't be combined.

$$4 + \sqrt{2}$$

So, in this next example, you could simplify the root, but that's the best you could do:

$$5 + \sqrt{8} = 5 + 2\sqrt{2}$$

Try some:

q)
$$12 + \sqrt{20}$$

r)
$$7 - \sqrt{60}$$

Finally, suppose that I was finding the slope intercept form of a line. Finish the work I've begun below by dividing the three into the slope and into the y-intercept.

$$2x + 3y = 9$$

$$3y = -2x + 9$$

$$y = \frac{-2x + 9}{3}$$

The same idea can follow with roots.

$$\frac{4 + 2\sqrt{3}}{2}$$

The 2 in the denominator must go into both terms of the numerator. (It can't go into just one!)

$$\frac{4 + 2\sqrt{3}}{2} = \frac{2 + \sqrt{3}}{1} = 2 + \sqrt{3}$$

Work these:

s) $\dfrac{8+4\sqrt{5}}{2} =$

t) $\dfrac{6-3\sqrt{7}}{3} =$

Finally, simplify the root. Then, if possible, divide.

u) $\dfrac{10+\sqrt{32}}{2} =$

v) $\dfrac{25-\sqrt{75}}{5} =$

Simplify the following: (It is fine to use a calculator, just notice that you should be entering them each in a slightly differently way.)

a) $\dfrac{\sqrt{9}}{\sqrt{16}} =$

b) $\sqrt{\dfrac{9}{16}} =$

c) $\dfrac{\sqrt{81}}{\sqrt{121}} =$

d) $\sqrt{\dfrac{81}{121}} =$

e) What did you find after working the two different versions of the same numbers?

Whenever it is convenient, we can switch between the two approaches. We can have a fraction under one square root or we can give the root to the numerator or denominator. Why would we want to? Under one square root, we could first reduce the fraction to make the work easier.

$$\sqrt{\frac{18}{50}} = \sqrt{\frac{2 \cdot 3 \cdot 3}{2 \cdot 5 \cdot 5}} = \sqrt{\frac{3 \cdot 3}{5 \cdot 5}} = \sqrt{\frac{9}{25}}$$

Now that it is simplified, the numerator and denominator become two simple square roots.

$$\frac{\sqrt{9}}{\sqrt{25}} = \frac{3}{5}$$

Try some. Start by simplifying the fraction and then separate them into two roots.

f) $\sqrt{\dfrac{20}{36}} =$

g) $\sqrt{\dfrac{54}{96}} =$

The idea is the same with variables. I'm going to simplify first.

$$\sqrt{\dfrac{x^4}{x^2}} = \sqrt{\dfrac{x \cdot x \cdot x \cdot x}{x \cdot x}} = \sqrt{x^2} = x$$

Try some:

h) $\sqrt{\dfrac{y^6}{y^2}} =$

i) $\sqrt{\dfrac{x^4 y^3}{x^2 y}} =$

And it is okay to have more left in the denominator:

$$\sqrt{\dfrac{x^2}{x^4}} = \sqrt{\dfrac{1}{x^2}} = \dfrac{1}{x}$$

Work these:

j) $\sqrt{\dfrac{y^3}{y^7}} =$

k) $\sqrt{\dfrac{x^6 y^2}{x^2 y^4}} =$

If you have both numbers and variables involved, simplify the numbers and variables individually.

$$\sqrt{\frac{8x^4}{50x^2}} = \sqrt{\frac{2 \cdot 2 \cdot 2 \cdot x \cdot x \cdot x \cdot x}{2 \cdot 5 \cdot 5 \cdot x \cdot x}} = \sqrt{\frac{2 \cdot 2 \cdot x \cdot x}{5 \cdot 5}} = \frac{2x}{5}$$

Try these:

l) $\sqrt{\dfrac{32a^5}{200a}} =$

m) $\sqrt{\dfrac{24x^8 y^2}{54x^4}} =$

Finally, it is possible that we aren't left with perfect squares. Simplify the fraction first, and then simplify the roots on the top and on the bottom as if they were two separate problems.

$$\frac{45x^2}{x^4} = \sqrt{\frac{5 \cdot 3 \cdot 3 \cdot x \cdot x}{x \cdot x \cdot x \cdot x}} = \sqrt{\frac{5 \cdot 3 \cdot 3}{x \cdot x}} = \frac{\sqrt{5 \cdot 3 \cdot 3}}{\sqrt{x \cdot x}} = \frac{3\sqrt{5}}{x}$$

Work these problems.

n) $\sqrt{\dfrac{20y^3}{y^5}} =$

o) $\sqrt{\dfrac{25a^3}{a}} =$

p) $\sqrt{\dfrac{18r^6s}{81r^3s^5}} =$

In this final problem, nothing in the fractions reduces, so separate it into two roots, one on top and one on bottom.

q) $\sqrt{\dfrac{9x^{10}}{16y^6}} =$

Algebra I

Active Lesson: 9.3a

Simplify the following by combining like terms:

a) $2x + 5 + 7x - 2$

b) $3y - 2x + 10 - 8x + 12 + y$

c) $15xy - 12xy + 2$

As I mentioned in the last activity, when it comes to combining like terms, roots act like variables. If the roots match, they can be combined.

$$3\sqrt{2} + 5\sqrt{2} = 8\sqrt{2}$$

Combine like terms:

d) $12\sqrt{3} - 10\sqrt{3}$

e) $6\sqrt{2} + 11\sqrt{2} - 5\sqrt{2}$

f) $7\sqrt{5} + 3\sqrt{7} - 2\sqrt{5} + 4\sqrt{7}$

It doesn't matter if the roots are numbers or variables, the idea remains the same.

$$2\sqrt{x} + 3\sqrt{x} = 5\sqrt{x}$$

Try some:

g) $7\sqrt{y} - 9\sqrt{y}$

h) $3\sqrt{v} - 2\sqrt{v} + 10\sqrt{v}$

i) $8\sqrt{xy} - 5\sqrt{xy}$

j) $16\sqrt{x} + 2 - 8\sqrt{x} + 8$

k) $14\sqrt{m} - 2\sqrt{n} + 4\sqrt{n} - 3\sqrt{m}$

l) $7\sqrt{x} - 7\sqrt{x}$

Here, we want to take the natural next step from the last activity. Sometimes we have roots that need to be simplified first. Then, we can combine them.

$$\sqrt{8} + \sqrt{18} = 2\sqrt{2} + 3\sqrt{2} = 5\sqrt{2}$$

Try these on your own.

a) $\sqrt{27} - \sqrt{12}$

b) $\sqrt{32} + \sqrt{8}$

Remember, if there is a number already in front of a root, multiply it by anything you pull out. Simplify the following:

c) $2\sqrt{20} + 3\sqrt{45}$

d) $17\sqrt{8} - 2\sqrt{72}$

Sometimes, not everything can combine.

e) $3\sqrt{75} - 5\sqrt{27} + \sqrt{50}$

As we saw previously, the ideas are the same with variables. But the way they combine needs some comment. Combine like terms.

f)
$$3xy + 2xy$$

The variables work like names, and the names must be identical in order to combine. Now look at this.

$$3x\sqrt{x} + 2x\sqrt{x}$$

The x is obviously a variable and the root works like a variable so to combine them the entire name must be the same. If they are, you can combine.

$$3x\sqrt{x} + 2x\sqrt{x} = 5x\sqrt{x}$$

Try some.

g) $5y\sqrt{y} - 3y\sqrt{y}$

h) $10x\sqrt{x} + 2x\sqrt{x} - 3x\sqrt{x}$

i) $7v\sqrt{v} - 4w\sqrt{w} + 2v\sqrt{v} + 6w\sqrt{w}$

Be careful with this next one.

j) $15x\sqrt{x} - 2x\sqrt{x} + 3\sqrt{x} + 7\sqrt{x} - 3$

Finally, sometimes the roots must be simplified before they can be combined. Try these.

k) $14\sqrt{x^3} - 2x\sqrt{x}$

l) $7\sqrt{x^3} + 5y\sqrt{y} - 4\sqrt{x^3} + 2\sqrt{y^3}$

m) $\sqrt{16x^3} - 2\sqrt{9x^3}$

n) $14y\sqrt{y} + 8\sqrt{y^3} - \sqrt{49y}$

o) $3x\sqrt{18} - \sqrt{8x^2}$

p) $x\sqrt{48} - 2\sqrt{3x^2} + 5\sqrt{75x^2}$

Algebra I

Active Lesson: 9.4a

Multiply the following:

a) $3x^2 \cdot 5x^3 =$

b) $7y^5 \cdot 4y^2 =$

When we solve problems like this, we are multiplying the things which are alike. We multiply the numbers and we multiply the variables. It is the same with roots. We multiply the numbers and we multiply the roots.

$$2\sqrt{3} \cdot 4\sqrt{5} = 2 \cdot 4\sqrt{3 \cdot 5} = 8\sqrt{15}$$

Try a couple:

c) $5\sqrt{2} \cdot 3\sqrt{7} =$

d) $15\sqrt{5} \cdot 3\sqrt{2} =$

Sometimes, when you multiply radicals, what you create can be simplified.

$$\sqrt{2} \cdot \sqrt{6} = \sqrt{12} = \sqrt{2 \cdot 2 \cdot 3} = 2\sqrt{3}$$

Multiply these and then simplify.

e) $\sqrt{6} \cdot \sqrt{3} =$

f) $\sqrt{10} \cdot \sqrt{2} =$

Sometimes you need to simplify and it joins what is already outside. (Remember, when it joins what is already outside, you multiply.)

$$3\sqrt{5} \cdot 2\sqrt{10} = 3 \cdot 2\sqrt{5 \cdot 10} = 6\sqrt{50} = 6\sqrt{2 \cdot 5 \cdot 5} = 6 \cdot 5\sqrt{2} = 30\sqrt{2}$$

Work these:

g) $7\sqrt{6} \cdot 5\sqrt{2} =$

h) $4\sqrt{15} \cdot 2\sqrt{3} =$

i) $3\sqrt{6x} \cdot 2\sqrt{3x} =$

In these next problems, the roots could be simplified from the beginning. However, it is probably easier to multiply everything and simplify at the end.

j) $5\sqrt{10x^2} \cdot 3\sqrt{8x^3} =$

k) $9\sqrt{2x^3y^2} \cdot 4\sqrt{20xy^3} =$

To see our final idea, multiply the following:

l) $\sqrt{2} \cdot \sqrt{2} =$

m) $\sqrt{9} \cdot \sqrt{9} =$

500

n) $\sqrt{x} \cdot \sqrt{x} =$

o)   What happens when you multiply a square root times itself?

Try these.

p) $4\sqrt{7} \cdot 3\sqrt{7} =$

q) $5\sqrt{x} \cdot 10\sqrt{x} =$

r) $\left(-3\sqrt{2}\right)^2 =$

s) $\left(7\sqrt{x}\right)^2 =$

Simplify the following using the distributive property.

a) $3(x - 5)$

b) $10(2y + 7)$

c) $x(3x - 4)$

d) $5y(6y + 9)$

As we often see, the concept can be extended; this time with roots. First, multiply the following:

e) $(7\sqrt{2}) \cdot 3 =$

f) $(7\sqrt{2}) \cdot 5\sqrt{6} =$

g) $(3\sqrt{3x}) \cdot (5\sqrt{6x^2}) =$

h) $(3\sqrt{3x}) \cdot 8 =$

Let's extend the multiplication of roots with the distributive property. Try these. (Hint: Look at the work you've already done.)

i) $7\sqrt{2}(3 + 5\sqrt{6}) =$

j) $3\sqrt{3x}(5\sqrt{6x^2} + 8) =$

Next, multiply the following.

k) $(x + 2)(2x - 3)$

l) $(x^2 - 5)(3x + 2)$

Now, multiply out the following:

m) $(2\sqrt{3}) \cdot 3 =$

n) $(2\sqrt{3}) \cdot (5\sqrt{2}) =$

o) $4 \cdot 3 =$

p) $4 \cdot (5\sqrt{2}) =$

Now, multiply out the following:

q)

$$(2\sqrt{3} + 4) \cdot (3 + 5\sqrt{2}) =$$

There are some special patterns which occur when we multiply. They follow the patterns we saw when we first multiplied binomials.

Multiply:

r) $(x + 6)(x + 6) =$

s) $(y - 5)^2 =$

In this pattern, we square the front term, square the back term, and then get two of the first term times the last. It follows with roots too. You don't have to memorize it. It will just occur when you multiply. Just be sure to combine like terms. Multiply these:

t) $(3\sqrt{2} + 6)(3\sqrt{2} + 6) =$

u) $(2 - 4\sqrt{3})^2 =$

We have one final pattern. This one will be very important when working with roots in the future. Again, it follows something we already know. Multiply these:

v) $(x + 9)(x - 9) =$

w) $(2y - 5)(2y + 5) =$

In this pattern, we square the first term, square the last term, and give the new binomial a negative sign in-between. Multiply these. Be sure to combine like terms.

x) $(7\sqrt{5} - 2)(7\sqrt{5} + 2) =$

y) $(3 + 4\sqrt{3})(3 - 4\sqrt{3}) =$

Notice that in these problems, the square roots have been eliminated. This trick will prove useful for just that reason.

505

Algebra I

Active Lesson: 9.5a

Earlier we saw that when we have a fraction under a square root, we get the same answer if we split the roots into the numerator and denominator.

$$\sqrt{\frac{9}{16}} = \frac{\sqrt{9}}{\sqrt{16}} = \frac{3}{4}$$

Since they are equal, we can go back and forth between a single root or the individual roots whenever it is helpful. We used this trick previously. Here we will first bring two roots back into one.

$$\frac{\sqrt{18}}{\sqrt{32}} = \sqrt{\frac{2 \cdot 3 \cdot 3}{2 \cdot 2 \cdot 2 \cdot 2 \cdot 2}}$$

This allows us to cancel. Then we can separate back into two roots.

$$\sqrt{\frac{3 \cdot 3}{2 \cdot 2 \cdot 2 \cdot 2}} = \frac{\sqrt{9}}{\sqrt{16}} = \frac{3}{4}$$

Try some:

a) $\dfrac{\sqrt{20}}{\sqrt{45}} =$

b) $\dfrac{\sqrt{98}}{\sqrt{128}} =$

The concept is the same with variables. Bring the two roots back into one so you can cancel.

c) $\dfrac{\sqrt{75x^5}}{\sqrt{3x}} =$

d) $\dfrac{\sqrt{20y}}{\sqrt{5y^3}} =$

The mistake students often make with these problems is failing to bring the two roots back together as one. It can be done without that step, but it leaves difficult and confusing variables to reduce. Try some more, being sure to first bring them back together under one root.

e) $\dfrac{\sqrt{15x^5}}{\sqrt{3x}} =$

f) $\dfrac{\sqrt{21x^9y}}{\sqrt{7xy^3}} =$

g) $\dfrac{\sqrt{200r^7s^2}}{\sqrt{r^3s^5}} =$

Algebra I

Active Lesson: 9.5b

Multiply the following:

a) $\sqrt{5} \cdot \sqrt{5} =$

b) $\sqrt{x} \cdot \sqrt{x} =$

In math, we never want to leave a radical in the denominator, so we do a trick.

$$\frac{1}{\sqrt{2}}$$

To get rid of the $\sqrt{2}$, we multiply this fraction by $\frac{\sqrt{2}}{\sqrt{2}}$.

$$\frac{1}{\sqrt{2}} \cdot \frac{\sqrt{2}}{\sqrt{2}}$$

This is legal because $\frac{\sqrt{2}}{\sqrt{2}} = 1$. Finish this problem by multiplying the numerators and the denominators.

c)
$$\frac{1}{\sqrt{2}} \cdot \frac{\sqrt{2}}{\sqrt{2}} =$$

This is called "rationalizing the denominator." Rationalize the denominator of the following. Be sure to show your work.

(Here you will multiply by $\frac{\sqrt{5}}{\sqrt{5}}$.)

d) $\frac{1}{\sqrt{5}}$

(Here you will multiply by $\frac{\sqrt{2}}{\sqrt{2}}$.)

e) $\dfrac{3}{\sqrt{2}}$

(Here you will multiply by $\frac{\sqrt{6}}{\sqrt{6}}$. After you rationalize the denominator, be sure to simplify the numerator.)

f) $\dfrac{\sqrt{3}}{\sqrt{6}}$

When a fraction is under a single root, we have learned that we can separate it into two. Finish this problem by simplifying the numerator and rationalizing the denominator.

g) $\sqrt{\dfrac{4}{5}} = \dfrac{\sqrt{4}}{\sqrt{5}} =$

Try these.

h) $\sqrt{\dfrac{16}{3}} =$

i) $\sqrt{\dfrac{9}{7}} =$

If the denominator can be simplified first, it should be. For instance:

$$\sqrt{\frac{3}{20}} = \frac{\sqrt{3}}{\sqrt{20}} = \frac{\sqrt{3}}{2\sqrt{5}}$$

Now, we will only rationalize with $\frac{\sqrt{5}}{\sqrt{5}}$. It will make the work much easier. Finish rationalizing:

j) $\dfrac{\sqrt{3}}{2\sqrt{5}} \cdot \dfrac{\sqrt{5}}{\sqrt{5}} =$

Work these:

k) $\dfrac{\sqrt{5}}{\sqrt{27}} =$

l) $\dfrac{\sqrt{7}}{\sqrt{32}} =$

m) $\dfrac{\sqrt{2}}{\sqrt{162}} =$

Multiply the following:

a) $(3 - 4\sqrt{2})(3 + 4\sqrt{2}) =$

b) $(5 + 2\sqrt{3})(5 - 2\sqrt{3}) =$

When we multiply and the only difference between the two terms is their sign, we are multiplying by conjugates. Give the conjugate of the following:

c) $(7 + 2\sqrt{15})$

d) $(\sqrt{x} - \sqrt{2})$

The value of conjugates involving roots is, that when multiplied, the roots are replaced with real numbers. This is the trick we want to use in order to rationalize a denominator that has two terms. We will multiply the top and bottom by the conjugate of the bottom. Finish the problem which I've started below.

e) $\dfrac{3}{2 + 2\sqrt{3}} \cdot \dfrac{(2 - 2\sqrt{3})}{(2 - 2\sqrt{3})} = \dfrac{3(2 - 2\sqrt{3})}{(2 + 2\sqrt{3})(2 - 2\sqrt{3})} =$

Try some on your own. Rationalize the denominator.

f) $\dfrac{2}{(3 - 4\sqrt{2})}$

g) $\dfrac{\sqrt{2}}{(5 + 2\sqrt{3})}$

h) $\dfrac{\sqrt{x}}{7 + 2\sqrt{15}}$

i) $\dfrac{\sqrt{x} + \sqrt{2}}{\sqrt{x} - \sqrt{2}}$

a)   What is the opposite of a square root?

Solve the following equation. With an equation, if you do something to one side, you must also do the same to the other.

b) $\sqrt{x} = 3$

Many types of equations in algebra follow the same pattern. Get the difficult portion of the equation alone. Then, once the difficult portion is alone, do its opposite to both sides of the equation. Look at this equation.

$$X + 4 = 10$$

c)   What would you do to get the X alone?

In this next equation, the $\sqrt{x}$ is like the X, get it alone. Then square both sides. Solve:

d) $\sqrt{x} + 4 = 10$

Next, look at this equation:

$$2X - 5 = 3$$

e)   What is the first step to get the X alone?

f)   What is the second step to get the X alone?

In this equation, the $\sqrt{z}$ is like the X. Get it alone. Then square both sides. Solve:

g) $2\sqrt{z} - 5 = 3$

That is the trick we will use to solving radical expressions. Isolate the radical, and then square both sides. Try a few harder versions.

Solve:

h) $\sqrt{x} - 4 = 12$

i) $\sqrt{3x - 3} = 3$

j) $\sqrt[3]{x - 2} = 2$ (Think this one through a bit. The trick is slightly different.)

k) $\sqrt{2x - 3} = \sqrt{3x - 4}$ (Here is a variation that is surprisingly easy. Just square both sides.)

The concept always stays the same, but the problems can get more difficult. On this next one, we need to square the entire right side, and that means foiling out a binomial. I've started it for you.

$$\sqrt{x + 1} + 1 = x$$
$$\sqrt{x + 1} = x - 1$$
$$x + 1 = (x - 1)^2$$
$$x + 1 = (x - 1)(x - 1)$$

Finish it from here. This will be quadratic. To solve it, you must get all the x's back together and make a room equal zero. (Here, the one on the left.) Then factor to solve.

l) $x + 1 = (x - 1)(x - 1)$

You should have gotten two answers. One of the answers was $x = 0$. However, that answer doesn't actually work. In the space below, plug 0 into the original equation and show that it doesn't work.

m)
$$\sqrt{x+1} + 1 = x$$

We have squared both sides to make the math easier, but doing this has changed the problem slightly. Let's see why.

At one point in the problem, we have $\sqrt{x+1} = x+1$. So, here's a trick to find the solution with a graph. If we had $y = \sqrt{x+1}$ and $y = x - 1$, the points where they intersect must be the answer.

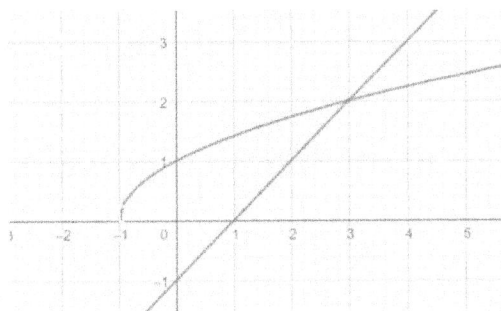

Notice that they hit when $x = 3$ but not $x = 0$.

Now I've graphed each of them after they've been squared.

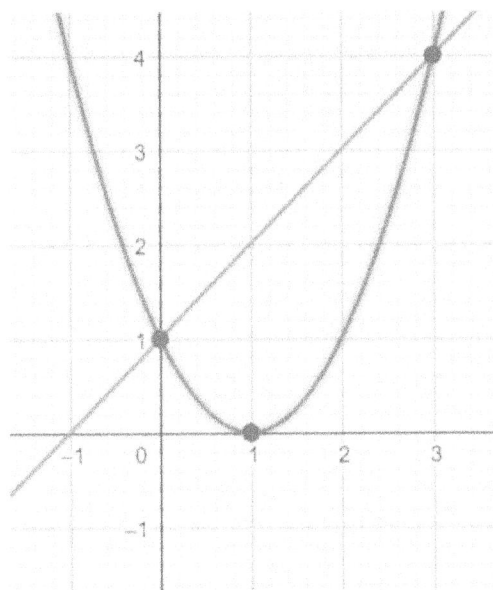

Notice they hit when $x = 3$ and $x = 0$.

So, what happened? When we squared both sides, we've created a quadratic equation that wasn't originally there. Here, the shape of the graph comes back up and hits a second time.

Here's the point. By squaring a root, it makes the math easier. But we've begun working with a quadratic which wasn't originally there. So, we need to check the answers. It is possible that a solution appears to be correct, but really isn't.

Try one final problem, and be sure to check your answers.

n) $\sqrt{x+4} - 2 = x$

Algebra I

Active Lesson: 9.6b

We have learned how to solve equations which involve square roots. Here, we are going to take a look at three applications which involve a square root formula.

The first idea is the area of a square. Since the sides of a square are equal, if we had the area and wanted to find a side, we could use this formula involving a square root.

$$s = \sqrt{A}$$

a) Here's a problem. A square patio is $400\ m^2$. What is the length of a side? We have the area, so use the formula to solve for s.

Try another problem where the length of the sides will involve a decimal. Round your answer to the nearest tenth.

b) The frame of square painting is $65\ in^2$. What is the length of a side?

If an object is dropped from a height, the time it takes to hit the ground can be approximated by this formula.

$$t = \frac{\sqrt{h}}{4}$$

Where h stands for the height from which the object is dropped and t stands for the time it takes to hit the ground. Try a problem.

c) A ball is dropped from a height of 250 feet. Approximate the time it will take to hit the ground. (Round your answer to the hundredths place.

Finally, the speed of a car can be estimated by the length of the skid marks it has left behind:

$$s = \sqrt{24d}$$

Where d stands for the length of the skid mark and s represents the speed of the car. Try this problem.

d) A skid mark left by a car is 200 feet long, estimate the speed of the car. (Round to the nearest tenth.)

Algebra I

Active Lesson: 9.7a

The topics in this chapter have previously been introduced. We will review them. Find the prime factorization of the following numbers.

a) 27

b) 8

c) 125

d) 16

e) 32

We are going to work higher order roots. (Roots other than square roots.) The number on the "house" is called the index. It tells us how many we need in order to bring a number outside. Here it takes three-of-a-kind to bring one out.

$$\sqrt[3]{27} = \sqrt[3]{3 \cdot 3 \cdot 3} = 3$$

Work these:

f) $\sqrt[3]{8} =$

g) $\sqrt[3]{125} =$

The idea extends to any index. This one requires four-of-a-kind.

h) $\sqrt[4]{16} =$

Here it requires five-of-a-kind.

i) $\sqrt[5]{32} =$

Remember, if we bring out multiple numbers, we multiply them back together.

$$\sqrt[3]{216} = \sqrt[3]{2 \cdot 2 \cdot 2 \cdot 3 \cdot 3 \cdot 3} = 2 \cdot 3 = 6$$

Try these:

j) $\sqrt[3]{64} =$

k) $\sqrt[4]{1296} =$

To see our next idea, multiply these:

l) $3 \cdot 3 =$

m) $-3 \cdot -3 =$

n) $-3 \cdot -3 \cdot -3 =$

Now, look at these roots. The first one can't be done. But there is an answer to the second. Provide the answer.

$\sqrt{-9} = Does\ Not\ Exist$

o) $\sqrt[3]{-27} =$

We can't have negatives under square roots (or any even-numbered index) because an even number of negative numbers always makes a positive. But we can have negatives under cube roots (or any odd-numbered index) because it can be done.

Circle any of the following which can be done. (You don't have to finish them.)

p) $\sqrt[4]{-256}$

q) $\sqrt[3]{-125}$

r) $\sqrt[7]{-78125}$

s) $\sqrt[6]{-64}$

Working higher-order roots with variables follows the same idea.

$\sqrt[3]{x^3} = \sqrt[3]{x \cdot x \cdot x} = x$

Work these:

t) $\sqrt[3]{y^3} =$

u) $\sqrt[4]{r^4} =$

v) $\sqrt[5]{z^5} =$

w) $\sqrt[3]{x^9} =$

x) $\sqrt[4]{y^8} =$

y) $\sqrt[3]{x^6 y^{12}} =$

As we saw before, we can work numbers and variables.

$\sqrt[3]{8x^3} = \sqrt[3]{2 \cdot 2 \cdot 2 \cdot x \cdot x \cdot x} = 2x$

Try some:

z) $\sqrt[3]{27y^3} =$

aa) $\sqrt[3]{125x^6 y^9} =$

This has a negative under the cube-root. Just work the numbers like normal, and because the negative is allowed, just add it to the outside.

bb) $\sqrt[3]{-8v^{12}} =$

cc) $\sqrt[4]{625x^{20}} =$

dd) $\sqrt[5]{243y^{10}x^5} =$

This last idea is one I don't ask students to do in the classroom. It involves adding an absolute value to variables which have come from under a root.

As we'll see later, there are two numbers which solve this problem: $\sqrt{9}$. It is both ± 3. However, for reasons we'll discuss in Algebra II, mathematicians typically want "the principal root," which refers to only the positive value. So, if a variable comes out from under a root, we don't know if it is a positive or negative number. But only the principal root is wanted. As a result, an absolute value is added to show that the positive value is the only one which we are interested in.

$$\sqrt{x^2} = |x|$$

There are no principal roots for odd roots. Any value is permitted to come out. Again, there is more to the discussion, but I will leave it for Algebra II.

Algebra I

Active Lesson: 9.7b

The idea in this activity is also one that was previously introduced. We are going to simplify higher-order roots where not everything comes out from beneath. First, do the prime factorization.

$$\sqrt[3]{16} = \sqrt[3]{2 \cdot 2 \cdot 2 \cdot 2}$$

Then circle those which can come out. It takes three-of-a-kind here. Notice a 2 gets left behind.

$$\sqrt[3]{\boxed{2 \cdot 2 \cdot 2} \cdot 2} = 2\sqrt[3]{2}$$

Work these:

a) $\sqrt[3]{54}$

On this next problem, notice that this takes four-of-a-kind.

b) $\sqrt[4]{80}$

This requires five-of-a-kind.

c) $\sqrt[5]{96}$

In these next problems, multiply any factors that remain in the house back together. For example:

$$\sqrt[3]{48} = \sqrt[3]{2 \cdot 2 \cdot 2 \cdot 2 \cdot 3} = 2\sqrt[3]{2 \cdot 3} = 2\sqrt[3]{6}$$

Try some.

d) $\sqrt[3]{80}$

e) $\sqrt[4]{486}$

527

And as we've seen before, we can extend the idea to variables.

$$\sqrt[3]{x^4} = x\sqrt[3]{x}$$

Simplify the following:

f) $\quad \sqrt[3]{y^7}$

g) $\quad \sqrt[4]{m^5}$

h) $\quad \sqrt[3]{x^4 y^7}$

Finally, we can add numbers to the variables. Work these:

i) $\quad \sqrt[3]{54r^5}$

j) $\quad \sqrt[4]{48m^7}$

k) $\quad \sqrt[3]{24x^3 y^{13}}$

In this section, if we have a fraction under a root, we can use the trick where we move between two roots or one, depending on if it helps us. This time, we are working with higher-order roots.

$$\frac{\sqrt[3]{x^4}}{\sqrt[3]{x}} = \sqrt[3]{\frac{x^4}{x}} = \sqrt[3]{x^3} = x$$

Work the following:

a) $\dfrac{\sqrt[3]{y^8}}{\sqrt[3]{y^2}}$

b) $\dfrac{\sqrt[4]{z^2}}{\sqrt[4]{z^6}}$

Here are a couple of problems that include numbers. First, make one root so you can simplify.

c) $\dfrac{\sqrt[3]{32y^9}}{\sqrt[3]{4y^3}}$

d) $\dfrac{\sqrt[4]{48z^9}}{\sqrt[4]{3z^5}}$

These problems will have factors left inside the roots.

e) $\dfrac{\sqrt[3]{54x^9}}{\sqrt[3]{x^4}}$

f) $\dfrac{\sqrt[4]{243z^9}}{\sqrt[4]{z^3}}$

In this next set, we start with one root, however nothing will reduce. So, just make it two roots and then simplify.

g) $\sqrt[3]{\dfrac{16x^4}{y^6}}$

h) $\sqrt[4]{\dfrac{r^3}{16s^8}}$

Algebra I

Active Lesson: 9.7d

Combine the following:

a) $2\sqrt{2} + 3\sqrt{2}$

b) $5\sqrt{x} - 2\sqrt{x}$

c) $12\sqrt{y} - 6\sqrt{x} + 5\sqrt{y} - 2\sqrt{x}$

d) $3x\sqrt{x} + 5x\sqrt{x}$

e) $14x\sqrt{x} - 2x\sqrt{x} + 5\sqrt{x}$

The concept is the same with higher order roots. Combine like terms:

f) $7\sqrt[3]{2} - 2\sqrt[3]{2}$

g) $8\sqrt[4]{3} + 2\sqrt[4]{3}$

h) $18\sqrt[3]{3x} - 3\sqrt[3]{3x}$

i) $7y\sqrt[4]{y} + y\sqrt[4]{y}$

In these problems, first simplify the roots in order to see if they can be combined.

j) $2\sqrt[3]{24} + 5\sqrt[3]{81}$

k) $5\sqrt[3]{x^4} - x\sqrt[3]{x}$

l) $8\sqrt[4]{y^6} + y\sqrt[4]{81y^2}$

Simplify one more, but remember the "name" portion must be the same. This one doesn't have an exact match, and therefore can't be combine.

m) $3\sqrt[3]{40x^7} + 2\sqrt[3]{135x^4}$

Algebra I

Active Lesson: 9.8a

Simplify the following:

a) $x^2 \cdot x^3 =$

b) $y^5 \cdot y^3 =$

When you multiply variables like this, what do you do with their exponents?

Follow that same logic to simplify these. (Don't over think it.)

c) $x^{\frac{1}{2}} \cdot x^{\frac{1}{2}} =$

d) $y^{\frac{1}{3}} \cdot y^{\frac{1}{3}} \cdot y^{\frac{1}{3}} =$

e) $m^{\frac{1}{4}} \cdot m^{\frac{1}{4}} \cdot m^{\frac{1}{4}} \cdot m^{\frac{1}{4}} =$

Now try these.

f) $\sqrt{x} \cdot \sqrt{x} =$

g) $\sqrt[3]{y} \cdot \sqrt[3]{y} \cdot \sqrt[3]{y} =$

h) $\sqrt[4]{m} \cdot \sqrt[4]{m} \cdot \sqrt[4]{m} \cdot \sqrt[4]{m} =$

There is a connection between roots and exponents which we've seen before. Roots can be written as exponents. Rewrite the following roots as exponents. (I've done the first one for you.)

i) $\sqrt{x} = x^{\frac{1}{2}}$

j) $\sqrt[3]{y} =$

k) $\sqrt[4]{m} =$

Simplify the following by changing the fractional exponents to roots. Since they involve numbers, an answer can be found. (Again, I've helped with the first one.)

l) $(36)^{\frac{1}{2}} = \sqrt{36} = 6$

m) $(8)^{\frac{1}{3}} =$

n) $(16)^{\frac{1}{4}} =$

Fractional exponents are roots. The index of the root is the denominator of the fraction. And when we make a root an exponent, all of the normal exponent rules apply. Here is a negative exponent. Exponent rules tell us that we can move a negative exponent to the denominator and it becomes positive.

$$16^{-\frac{1}{2}} = \frac{1}{16^{\frac{1}{2}}} = \frac{1}{\sqrt{16}} = \frac{1}{4}$$

Try some:

o) $64^{-\frac{1}{2}}$

p) $16^{-\frac{1}{4}}$

q) $27^{-\frac{1}{3}}$

To see our next idea, simplify the following:

r) $(x^2)^3 =$

Follow the same idea to simplify this: (The secret here is: $\frac{1}{2} \cdot 3$)

s) $\left(x^{\frac{1}{2}}\right)^3 =$

t) Which of the following must be the same as $x^{\frac{5}{2}}$? (Hint: Two of them are correct.)

   a) $\sqrt[5]{x^2}$
   b) $\left(\sqrt{x}\right)^5$
   c) $\sqrt{x^5}$

Write the following roots as exponents. (There is nothing to solve.)

u) $\sqrt[3]{y^4}$

v) $\left(\sqrt[4]{r}\right)^3$

Write this expression with a fractional exponent. (Notice that the entire contents of the root are being squared.)

w) $\left(\sqrt[3]{2x}\right)^2$

Write this expression with a fractional exponent. (Here, everything is being taken to the 5th power.)

x) $\left(\sqrt{\dfrac{3x}{2}}\right)^5$

Write the following exponents as roots.

y) $\quad x^{\frac{2}{5}}$

z) $\quad y^{\frac{4}{3}}$

aa) $\quad (5z)^{\frac{2}{3}}$

bb) $\quad \left(\frac{4x}{5}\right)^{\frac{4}{5}}$

This last section involves real values and not variables. Therefore, we can simplify. When I do these, I take the roots first, so that the numbers are easier to work with.

$$9^{\frac{3}{2}} = \left(\sqrt{9}\right)^{3} = 3^3 = 27$$

Finish these:

cc) $\quad 4^{\frac{5}{2}}$

dd) $\quad 27^{\frac{2}{3}}$

ee) $\quad 9^{-\frac{3}{2}}$

This final problem doesn't exist, explain why:

ff) $\quad (-16)^{\frac{3}{2}}$

Algebra I

Active Lesson: 9.8b

Write the following roots as exponents.

a) $\sqrt{x^3}$

b) $\left(\sqrt[3]{y}\right)^5$

When we write roots as exponents, all of the exponent rules apply. We'll refresh ourselves on each of the exponent rules and then apply them to fractional exponents.

Simply.

c) $x^2 \cdot x^3 =$

d) $y^4 \cdot y^5 =$

e) What is the rule for exponents with like bases?

Now try the idea with fractions. It follows the same principal.

f) $x^{\frac{1}{3}} \cdot x^{\frac{1}{3}} =$

g) $z^{\frac{1}{2}} \cdot z^{\frac{1}{3}} =$

h) $y^{\frac{2}{3}} \cdot y^{\frac{1}{4}} =$

Next, let's refresh a power to a power. Simplify.

i) $(x^4)^3 =$

j) $(z^6)^2 =$

k)   What do we do when taking an exponent to an exponent?

Now, follow the same principal when working with fractional exponents.

l) $(x^3)^{\frac{1}{2}} =$

m) $(r^6)^{\frac{1}{3}} =$

n) $(m^6)^{\frac{3}{4}} =$

Next, let's review division. Simplify these:

o) $\dfrac{x^5}{x^3} =$

p) $\dfrac{y^{12}}{y^4} =$

Try the idea with fractions. Subtract the bottom fraction from the top fraction.

q) $\dfrac{x^{\frac{2}{3}}}{x^{\frac{1}{3}}} =$

r) $\dfrac{y^{\frac{3}{5}}}{y^{\frac{2}{5}}} =$

On this next idea, recall when an entire group is raised to an exponent, we give the exponent to everything inside. Try these:

s) $(2x)^3$

t) $(5y^2)^2$

u) $(x^3y^4)^5$

Again, the idea holds for fractional exponents.

v) $(9x^{\frac{2}{3}})^{\frac{1}{2}}$

w) $(8y^{\frac{1}{4}})^{\frac{2}{3}}$

x) $(r^2s^3)^{\frac{3}{4}}$

Next, we recall that negative exponents can be moved across the fraction line. Write the following with positive exponents.

y) $x^{-2} =$

z) $\frac{1}{y^{-3}} =$

aa) $\frac{v^2}{v^{-3}} =$

Work some problems which involve fractional exponents.

bb) $x^{-\frac{1}{2}} =$

cc) $\dfrac{1}{y^{-\frac{1}{3}}} =$

dd) $\dfrac{v^{\frac{2}{3}}}{v^{-\frac{2}{3}}} =$

Finally, we can use multiple exponential rules to simplify. (There are multiple ways to do this, but the easiest would be to move everything to the top and simply add.)

ee) $\dfrac{x^2 x^{-3}}{x^{-5}}$

ff) $\dfrac{y^{-4} y^{-3}}{y^{-12}}$

Extend this idea to fractional exponents.

gg) $\dfrac{x^{\frac{1}{2}} \cdot x^{-\frac{3}{2}}}{x^{-\frac{5}{2}}}$

hh) $\dfrac{y^{\frac{4}{3}} \cdot y^{-\frac{1}{3}}}{y^{-\frac{12}{3}}}$

a) What is the opposite of a square?

Solve the following equation. With an equation, if you do something to one side, you must also do the same to the other. And remember, a square root creates both $\pm$.

b) $x^2 = 64$

Many types of equations in algebra follow the same pattern. Get the difficult portion of the equation alone. Then, once the difficult portion is alone, do its opposite to both sides of the equation. Look at this equation.

$$X + 4 = 40$$

c) What would you do to get the X alone?

In this next equation, the y^2 is like the X, get it alone. Then take the square root of both sides. (Again, don't forget, a square root makes both $\pm$. Solve:

d) $$y^2 + 4 = 40$$

Next, look at this equation:

$$2X - 3 = 239$$

e) What is the first step to get the X alone?

f) What is the second step to get the X alone?

g) In this equation, the z^2 is like the X, get it alone. Then take the square root of both sides. Solve:

$$2z^2 - 3 = 239$$

The next few look a bit different, but notice that the same trick still applies. Everything on the left is still being squared. There is one other twist at the end, so let me show you one.

$$(x - 2)^2 = 16$$
$$\sqrt{(x - 2)^2} = \sqrt{16}$$
$$x - 2 = \pm 4$$
$$x = 2 \pm 4$$
$$x = 6 \ or \ -2$$

h) Explain how $x = 2 \pm 4$ turns into $x = 6 \ or \ -2$.

Try these on your own:

i) $(x + 3)^2 = 4$

j) $(y - 5)^2 = 9$

542

Algebra I

Active Lesson: 10.1b

Combine like terms:

a) $3 + 2x + 5 - 6x$

Now, combine these like terms:

b) $3 + 2\sqrt{2} + 5 - 6\sqrt{2}$

Recall that when combining like terms, roots work like variables. You can only combine them if you have another root of the same type.

Let's return to what we worked on in our last activity. In this problem, when you take the square root of the right side, it isn't a perfect square. So, we simplify the best we can.

$$(m - 3)^2 = 18$$
$$m - 3 = \sqrt{18}$$
$$m - 3 = \pm 3\sqrt{2}$$
$$m = 3 \pm 3\sqrt{2}$$

This is really two answers: $m = 3 + 3\sqrt{2}$ or $m = 3 - 3\sqrt{2}$. But the terms can't be combined, so this is the best we can do. Try a similar problem on your own.

c) $(x + 5)^2 = 40$

Solve the following problem:

d) $X^2 + 3 = 19$

Now solve this problem. First get $(b - 5)^2$ alone, take the square root, and then send the 5 to the other side. (The five will not combine with the root.)

e) $(b - 5)^2 + 6 = 19$

Try another:

f) $(y + 3)^2 - 6 = 46$

Next solve this:

g) $2X^2 - 4 = 20$

Follow the same idea to solve this:

h) $2(x + 5)^2 - 4 = 20$

Factor the following polynomials:

a) $x^2 + 6x + 9$

b) $y^2 - 18y + 81$

c) $z^2 - 10z + 25$

d) All of these are the same type of factoring problem called a Perfect Square Trinomial. Describe what they have in common when they are in factored form.

Look at the square roots of the last terms and the coefficients of the middle terms.

	$x^2 + 6x + 9$	$y^2 - 18y + 81$	$z^2 - 10z + 25$
Square root of last:	3	9	5
Coefficient of the middle:	6	18	10

e) What's the connection?

In this activity, we want to make a perfect square trinomial. (This idea is called Completing the Square.) To do that we need to find the missing final coefficient. The trick is to go in the other direction. First, we need half the middle term. Then we need to square it.

In these next problems, there is a number missing which would make them each perfect square trinomials, add that missing number. (Half the missing term. Square it.)

f) $x^2 + 22x +$ ____

g) $y^2 - 14y +$ ____

h) $z^2 - 24z +$ ____

To fully understand completing the square, we need to look back at some ideas from fractions. Divide the following fraction in half. (You can probably do this in your head, but I've given you a hint.)

$$\frac{1}{2} = \frac{\frac{1}{2}}{2} = \frac{\frac{1}{2}}{\frac{2}{1}} = \frac{1}{2} \cdot \frac{1}{2} =$$

Try these. Divide the fraction in half.

i) $\dfrac{\frac{1}{4}}{2} =$

j) $\dfrac{\frac{2}{3}}{2} =$

Next, we need to square a fraction. Remember, you can simply square the numerator and the denominator separately.

k) $$\left(\frac{1}{2}\right)^2 = \frac{(1)^2}{(2)^2} =$$

Work a few more. Square the following fractions.

l) $\left(\frac{3}{4}\right)^2$

m) $\left(\frac{5}{7}\right)^2 =$

Unfortunately, sometimes completing the square can involve fractions, but the process is still the same. Divide the middle number in half. Square that number. Find the missing number that would complete the square.

n) $x^2 + \dfrac{1}{2}x + \underline{\quad}$

o) $y^2 + \dfrac{2}{3}y + \underline{\quad}$

p) $r^2 - \dfrac{1}{4}r + \underline{\quad}$

546

Completing the square is a helpful math trick because it allows us to easily solve quadratic equations using the square root property. Here's a refresher on how. The below equation has a perfect square trinomial on the left.

$$x^2 + 6x + 9 = 16$$

$$(x + 3)^2 = 16$$

Square root both sides. (Remember, the square root of 16 is ± 4.)

$$x + 3 = \pm 4$$

$$x = -3 \pm 4$$

$$x = 1 \text{ or } x = -7$$

Solve the following quadratic equations:

a) $x^2 + 12x + 36 = 9$

b) $y^2 + 16y + 64 = 36$

Simplify the following square roots.

c) $\sqrt{50}$

d) $\sqrt{20}$

e) Can the following be simplified any further? Why or why not?

$$7 \pm 5\sqrt{2}$$

Sometimes the right side isn't a perfect square. Factor the right side and then take the square root of both sides. Simplify the left side the best you can and then solve for the variable.

f) $y^2 - 14y + 49 = 50$

g) $z^2 - 20z + 100 = 18$

Now we are ready to put all the ideas together in order to complete the square. To complete the square on the left we need 25. But this is an equation, and so we can't just add 25 to one side because we want to. We must add it to the other as well.

h) $x^2 - 10x + \underline{\quad} = 11 + \underline{\quad}$

$$x^2 - 10x + 25 = 11 + 25$$

Finish the problem below, solving for x:

i) $x^2 - 10x + 25 = 36$

Try some on your own. Complete the square and then solve.

j) $x^2 + 12x + \underline{\quad} = -27 + \underline{\quad}$

k) $y^2 + 16y + \underline{\quad} = -28 + \underline{\quad}$

l) $y^2 - 14y + \underline{\quad} = 1 + \underline{\quad}$

Sometimes we want to complete the square but there is an extra number in the way.

$$x^2 + 4x + 2 = 7$$

We just move it to the other side.

$$x^2 + 4x + \underline{\quad} = 7 - 2$$

$$x^2 + 4x + \underline{\quad} = 5$$

From here, the process is the same. Finish this problem:

m) $x^2 + 4x + \underline{\quad} = 5$

Try these. Begin by moving the extra number on the left over to the right so that you can complete the square.

n) $y^2 - 12y - 4 = 9$

o) $z^2 - 2z - 10 = 33$

So far, as we completed the square, the leading term has always been a 1. But what do we do if it isn't.

$$2x^2 + 4x + 10 = 40$$

There is one extra step. We can make it a 1 by dividing each term in the equation by that leading term.

$$\frac{2x^2}{2} + \frac{4x}{2} + \frac{10}{2} = \frac{40}{2}$$

Finish completing the square just as we did previously.

a)
$$x^2 + 2x + 5 = 20$$

Try some on your own.

b) $2x^2 - 20x + 30 = 180$

c) $3z^2 - 18z - 9 = 18$

Finally, clearing that leading term doesn't always work out so nicely. It is common for it to make fractions. Although in these cases, our next idea, called the quadratic formula, is probably easier. But let's try by completing the square.

$$2x^2 - 6x - 4 = 2$$

$$\frac{2x^2}{2} - \frac{6x}{2} - \frac{4}{2} = \frac{2}{2}$$

$$x^2 - 3x - 2 = 1$$

$$x^2 - 3x + \underline{\quad} = 3$$

$$x^2 - 3x + \frac{9}{4} = 3 + \frac{9}{4}$$

$$x^2 - 3x + \frac{9}{4} = \frac{12}{4} + \frac{9}{4}$$

$$\left(x - \frac{3}{2}\right)^2 = \frac{21}{4}$$

$$x - \frac{3}{2} = \pm\sqrt{\frac{21}{4}}$$

$$x = \frac{3}{2} \pm \frac{\sqrt{21}}{\sqrt{4}} = \frac{3}{2} \pm \frac{\sqrt{21}}{2}$$

Try one on your own.

d) $3y^2 - 2y + 3 = 12$

Algebra I

Active Lesson: 10.3a

In the last few activities, we've been finding the solutions to quadratic equations. These solutions have many names. They can also be referred to as roots or x-intercepts. All of these terms are synonyms. Here is a graph of the quadratic equation $2x^2 + 2x - 4$.

a)

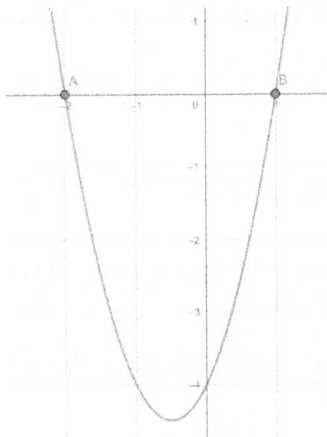

What are the coordinates of the x-intercepts? (Remember, intercepts are points, so please give them as an ordered pair.)

There are several ways to find the solutions (x-intercepts, roots) of a quadratic equation. However, one that will always work is called the quadratic formula. It is found by completing the square of a generic quadratic equation. Here is the formula:

$$x = \frac{-b \pm \sqrt{b^2 - 4ac}}{2a}$$

The $a, b,$ and c are taken from a quadratic equation in standard form:

$$ax^2 + bx + c = 0$$

The a is the number in front of the x^2. The b is the number in front of the x. And, the c is the number by itself.

Use the quadratic formula to find the x-intercepts for $2x^2 + 2x - 4$. The answers should match what you found on the graph. However, you must show your work. (Remember the order of operations. You must start by cleaning up what is under the square root.) I've set up the equation for you. Be careful with the signs. (You should get 36 under the root.)

$$2x^2 + 2x - 4$$

b)

$$x = \frac{-(2) \pm \sqrt{(2)^2 - 4(2)(-4)}}{2(2)}$$

553

c)   If done correctly, your solutions to the quadratic equation should be a match with your x-intercepts from the graph. Do they match?

Here is the graph of the quadratic $3x^2 + 3x - 6$.

d)   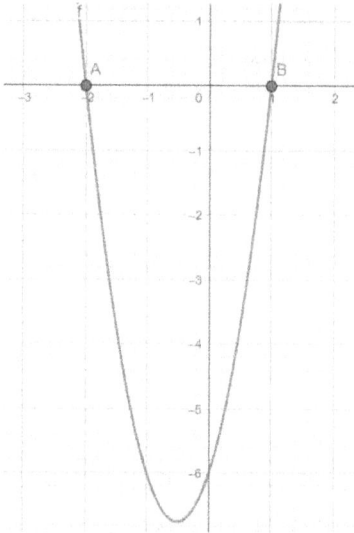   What are the coordinates of the x-intercepts?

e)   Use the quadratic formula to find the x-intercepts. Show your work to see that you get the x-intercepts from the graph. You are solving the equation $3x^2 + 3x - 6 = 0$.

Sometimes the solutions to a quadratic equation aren't so easy. Here is the graph of $2x^2 - 5x + 1$.

f)   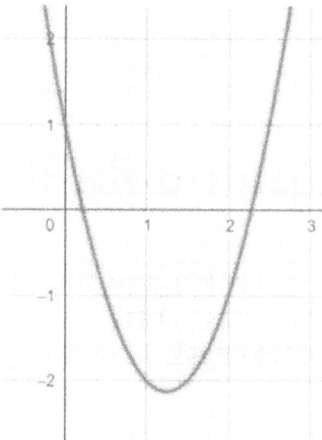   The solutions can't be read off the graph and are actually irrational numbers. Use the quadratic equation to find the solutions.

As you will learn more about in Algebra II, quadratic equations must have two solutions. So far that has been true, but look at the graph for the equation $3x^2 + 4x + 2$.

g)

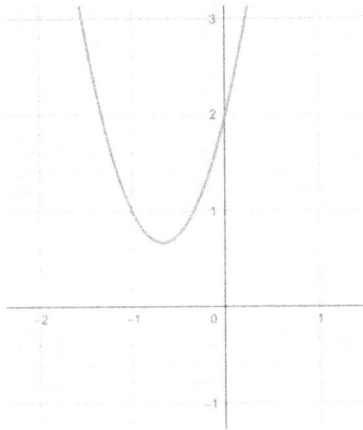

If x-intercepts are solutions, what is the problem here?

h)   To understand what's happening, we need to take a look at a peculiar square root. What is the problem here?

$$\sqrt{-4}$$

Despite the problem, mathematicians wondered what would happen if they continued to work with roots like this. And even though they didn't fit into the real world, the math could go on. To simplify working with these roots, they abbreviated $\sqrt{-1}$ into the term i. And i was called an imaginary number. So, the negative under a root became i and the rest of the root was worked as always.

$$\sqrt{-4} = \sqrt{4 \cdot -1} = \sqrt{4}i = \pm 2i$$

Use i to complete the following roots:

i) $\sqrt{-16}$

j) $\sqrt{-25}$

k) $\sqrt{-50}$

555

l)   Now back to the question, where are the roots in the equation $3x^2 + 4x + 2 = 0$? Use the quadratic formula to find the solutions. (Use imaginary numbers to simplify.)

There must be two solutions to a quadratic equation. So, where are the solutions to this equation? They involve imaginary numbers. And since imaginary numbers aren't real, the solutions don't touch the x-axis. Try another. Use the quadratic formula to find the solutions.

m)

$$2x^2 + 3x + 4 = 0$$

Questions will often simply say, solve using the quadratic formula. But they must have a zero on the right side. If they don't, we first need to rearrange. Solve these equations by getting the quadratic equations in standard form and then using the quadratic formula.

n) $4x^2 - 3x = -2$

This final problem requires you to first multiply out the left and then get the equation equal to zero.

o) $(x - 5)(x + 4) = 2$

Algebra I

Active Lesson: 10.3b

Solve the following problems using the quadratic formula:

$$x = \frac{-b \pm \sqrt{b^2 - 4ac}}{2a}$$

a) $3x^2 + x - 2 = 0$

b) $x^2 - 2x + 1 = 0$

c) $2x^2 - 5x + 9 = 0$

The portion of the quadratic formula under the square root is called the discriminant.

$$x = \frac{-b \pm \sqrt{b^2 - 4ac}}{2a}$$

It is useful because it will tell us: if we get two real solutions, if we get two imaginary solutions, or if we get one real solution. Find the value of the discriminant on the three problems you already solved.

$$b^2 - 4ac$$

d) $3x^2 + x - 2 = 0$

$$discriminant =$$

e) $x^2 - 2x + 1 = 0$

$$discriminant =$$

f) $2x^2 - 5x + 9 = 0$

$$discriminant =$$

The reason the discriminant gives us this information is just a bit of logic, since it is the value under the square root. Look at your previous answers and their discriminants. Then, match the following:

g)   A discriminant greater than zero.   One real solution

h)   A discriminant equal to zero.   Two imaginary solutions

i)   A discriminant less than zero.   Two real solutions

Use the discriminant to tell the types of solutions you would get to the following quadratic equations. You don't need to find the solutions, simply use the discriminant to reason out whether you would get: one real solution, two imaginary solutions, or two real solutions.

$$b^2 - 4ac$$

(If it creates a negative number, it will make two imaginary solutions. If it creates a zero, it will create one real solution. And, if it creates a positive number, it will make two real solutions.)

j) $3x^2 - 5x + 1$

k) $2x^2 + 5x + 6$

l) $x^2 + 4x + 4$

The word problems in this section are of a type we have already seen earlier in the course. Quadratic equations will be created. Previously, the equations could be solved by factoring. Now, we will need to solve using the quadratic formula.

Let's look at one.

The product of two consecutive integers is 210. Find the numbers.

As before, we start with our dictionary.

Math	English
x	1st Number
x+1	2nd Number

Building our equation, we would have:

$$x(x + 1) = 210$$

To use the quadratic formula, we need to clean up and get a zero on the right side.

$$x^2 + x = 210$$

$$x^2 + x - 210 = 0$$

a) Use the quadratic formula to finish finding the values of x.

Let's look at another.

A triangle has an area of 155 m^2. The base is 6 more than twice the height. Find the height of the triangle.

We are working with the area of a triangle, so our formula will be based on:

$$A = \frac{1}{2}b \cdot h$$

Here is our dictionary:

Math	English
x	height
2x+6	base

Setting up our equation we get:

$$155 = \frac{1}{2}(x)(2x + 6)$$

b) Simplify and arrange it for using the quadratic formula.

c) Use the quadratic formula to find the value of the height.

Next, we will use the Pythagorean theorem.

The legs of a right triangle are the same length. If the hypotenuse is 22 feet long, find the length of the legs.

The Pythagorean Theorem gives us:

$$a^2 + b^2 = c^2$$

And since the legs are the same length, we get:

$$x^2 + x^2 = 22^2$$

Or:

$$2x^2 = 484$$

d) Use the quadratic equation to solve for the length of the legs.

For the final problem, we will use the area of a rectangle.

The area of a rectangle is $55m^2$. If the length of the rectangle is two less than twice the width, find the length and the width.

Here is our dictionary.

Math	English
2x-2	Length
x	Width

The area of a rectangle is:

$$A = l \cdot w$$

e) Set up and solve for the length and the width. (Your answer will be the width. Use your dictionary to find the length.)

Algebra I

Active Lesson: 10.5a

We want to begin taking a detailed look at the graphs of quadratic equations. Quadratic equations look like this:

$$y = ax^2 + bx + c$$

The shape of a quadratic equation is a parabola:

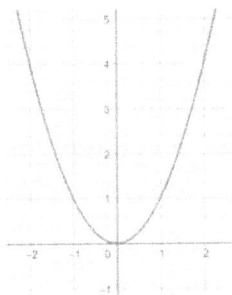

The vertex of a quadratic equation is the bottom of the U. In the picture above, the vertex is at $(0,0)$. But the vertex can be moved. Give the coordinates of the vertex for these parabolas:

a)

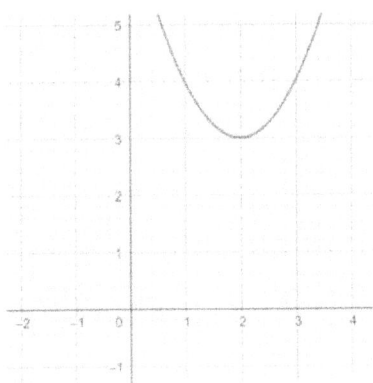

b)

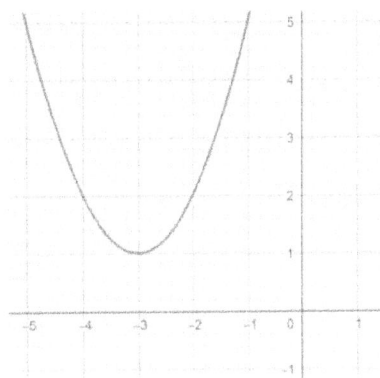

Vertex: Vertex:

With the help of Geogebra, we want to see some connections between the formula of a quadratic equation and its graph.

c) Below are the graphs of four quadratic equations. One thing in their formula is determining if the graphs are right side up or upside down. What is it?

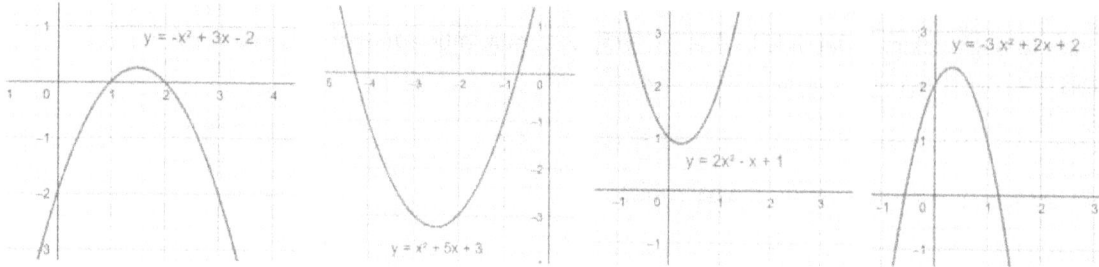

$y = -x^2 + 3x - 2$

$y = x^2 + 5x + 3$

$y = 2x^2 - x + 1$

$y = -3x^2 + 2x + 2$

For each of the following quadratic equations, list the vertex and solve this equation: $x = -\dfrac{b}{2a}$. The a is the first coefficient (before the x^2) and b is the second coefficient (before the x).

d)

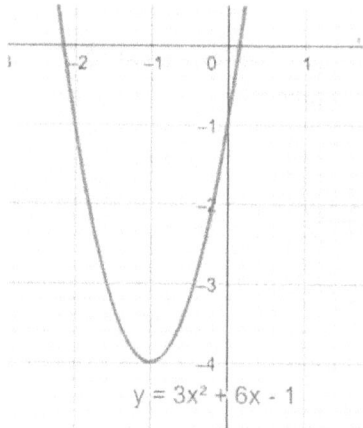

$y = 3x^2 + 6x - 1$

e)

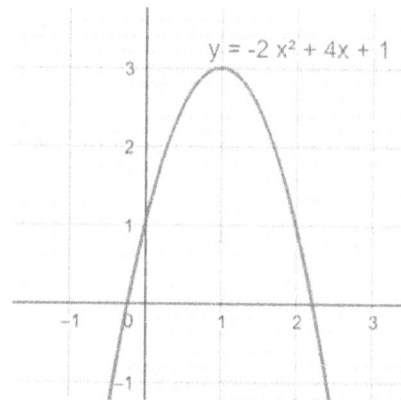

$y = -2x^2 + 4x + 1$

Vertex:

$x = -\dfrac{b}{2a} =$

Vertex:

$x = -\dfrac{b}{2a} =$

f) What is the connection between the vertex and the value of x from the formula?

The formula $x = -\frac{b}{2a}$ creates a line called the axis of symmetry. Quadratic equations are symmetrical (the same on either side) and the axis of symmetry is the line over which the equation can fold. The axis of symmetry is a vertical line. Give the equation for the axis of symmetry from the following graphs:

g)

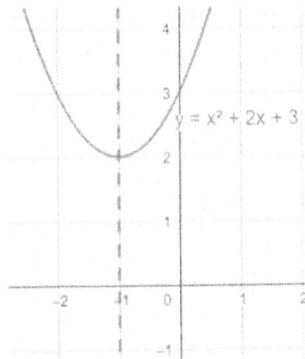

$y = x^2 + 2x + 3$

Vertex:

$x =$

h)

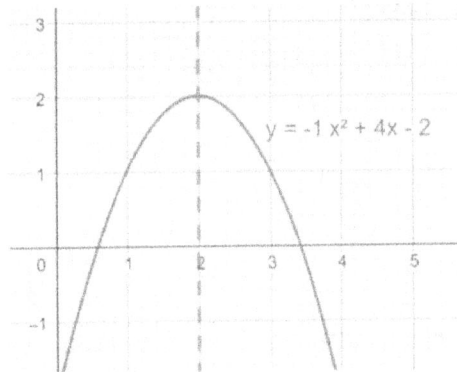

$y = -1 x^2 + 4x - 2$

Vertex:

$x =$

The axis of symmetry is a line, but plug that value of x into the equation.

$y = x^2 + 2x + 3$

$y = -x^2 + 4x - 2$

i) What is the connection between the vertex and the answer you get when you put the x value into the equation?

j) Find the vertex of the following quadratic equation without graphing:

$$y = 2x^2 - 4x + 2$$

Use the formula $x = -\frac{b}{2a}$ to find the x coordinate of the vertex. Then plug that value of x into the quadratic equation to find the y coordinate of the vertex.

Next, we want to find the x and y intercepts of a quadratic function. Just as with a line, the x-intercept happens when the value of y is zero.

$$y = x^2 - 3x - 4$$

$$0 = x^2 - 3x - 4$$

k) To find the x-intercepts, (there are two), factor and use the zero-product property.

$$x^2 - 3x - 4 = 0$$

l) Intercepts are points. Give the value of your two x-intercepts:

(,0)   (,0)

To find a y-intercept, it occurs when x is equal to zero. Substitute 0 in for x and find the value of the y-intercept:

$$y = x^2 - 3x - 4$$
$$y = (0)^2 - 3(0) - 4$$

m) A y-intercept is a point. (There is only one.) Give the value of the y-intercept:

(0,)

n) Try one on your own. Find the values of the x and y-intercepts for this quadratic equation:

$$y = x^2 + 8x + 7$$

Sometimes a quadratic equation does not have any x-intercepts. Look at this quadratic equation:

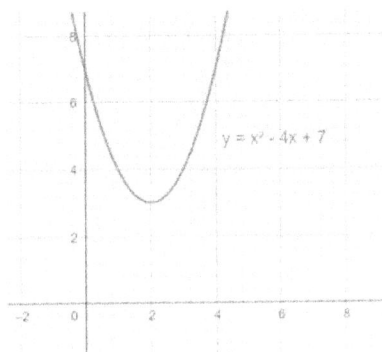

$y = x^2 - 4x + 7$

o) Solving a quadratic equation is the same thing as finding the x value of the x-intercept. Use the quadratic formula to discover why this quadratic doesn't have intercepts:

$$y = x^2 - 4x + 7$$

p) What happened when using the quadratic formula? Why does this explain why the quadratic equation doesn't have intercepts?

Algebra I

Active Lesson: 10.5b

Next, we want to graph quadratic equations. Essentially, we need the ideas we learned in the last activity. Let's graph this quadratic equation.

$$y = x^2 + 4x + 3$$

a) First, will this graph upward ($\cup$) or downward ($\cap$)?

b) Second, find the vertex. (Remember to plug the value of the axis of symmetry into the equation to find the y-coordinate of the vertex.)

$$x = -\frac{b}{2a}$$

Third, find the x-intercepts by factoring and the zero-product property. (This is when $y = 0$) There will be two. Give your answers as points.

c)
$$x^2 + 4x + 3 = 0$$

Find the y-intercept. This is when we substitute zero in for x. Give your answer as a point.

d)
$$y = (0)^2 + 4(0) + 3$$

571

Finally, use this information to graph the equation. Mark your vertex and your intercepts. Knowing the basic shape of a quadratic function, and that it is symmetrical, you can then finish the graph.

e)

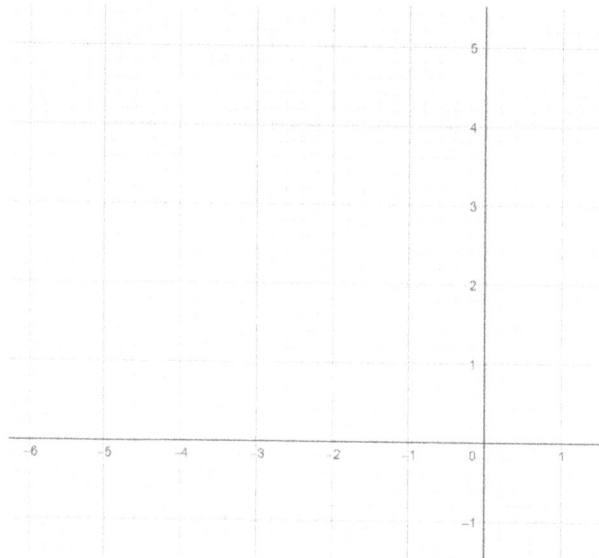

Try another:

Graph the quadratic equation.

$$y = x^2 + 6x + 8$$

f) Upward or downward?

g) Vertex:

h) x-intercept:

i) y-intercept:

j)

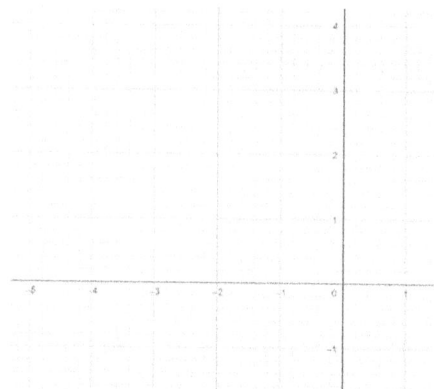

Finally, we want to graph a quadratic equation which doesn't have x-intercepts.

$$y = -x^2 + 4x - 5$$

k) Upward or downward?

l) Find the vertex.

m) There are no x-intercepts. (There couldn't be because of the location of the vertex and the direction that the parabola opens.) But we can find the y-intercept. Give the value of the y-intercept as a point.

n) Graph the information we currently have: vertex and y-intercept.

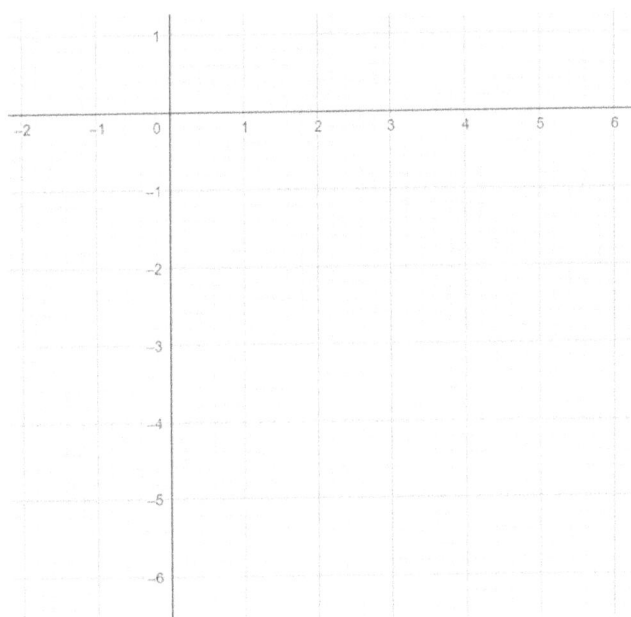

Add the axis of symmetry to the graph as a dashed vertical line. Since parabolas are symmetrical, add a point to the graph that is symmetrical to the y-intercept. (It will be the same number of horizontal steps away from the axis of symmetry, yet in the other direction.)

Finally, using the symmetrical point, graph your quadratic equation.

Algebra I

Active Lesson: 10.5c

In this activity, we are going to look at word problems related to quadratic equations. Among all the types of word problems you can come across, these are among the easiest.

A rocket is fired into the air. The height of the rocket is represented by this equation:

$h(t) = -3t^2 + 150t + 15.$

Find the maximum height of the rocket.

a) Look at the equation that they have given you. Is this quadratic equation right-side up (a U) or upside-down (a ∩)?

b) Find the value of the vertex. Remember, $h = -\frac{b}{2a}$ and $k = f(h)$.

The word problem asked you to find the maximum height of the rocket. You've already found it.

$$h = the\ time\ the\ rocket\ is\ at\ its\ maximum\ height.$$

$$k = the\ maximum\ height\ of\ the\ rocket.$$

c) What is the maximum height of the rocket?

Try another.

A football is thrown in the air from the top of a house. The height of the football is explained by this function: $h(t) = -5t^2 + 32t + 12$. Find the maximum height of the ball *and* the time it takes to reach this height.

d)   Is this quadratic equation right-side up (a U) or upside-down (a ∩)?

e)   Find the value of the vertex. Remember, $h = -\frac{b}{2a}$ and $k = f(h)$.

f)   What is the maximum height of the ball?

g)   How do we know that the vertex is the maximum and not the minimum?

h)   At what time does the ball reach the maximum height?

Algebra I

Answer Key

1.1a

a) Million

b) Hundred Thousand

c) Hundred

d) Ten Million

e) Hundred Million

f)

Trillion			Billion			Million			Thousand			Ones		
Hundred Trillion	Ten Trillion	Trillion	Hundred Billion	Ten Billion	Billion	Hundred Million	Ten Million	Million	Hundred Thousand	Ten Thousand	Thousand	Hundred	Tens	Ones
7	0	1	2	3	7	9	4	1	0	0	2	1	4	9

g) one hundred seventy-four

h) two hundred thirty-four

i) five hundred forty-nine

j) one hundred seventy-four million, two hundred thirty-four thousand, five hundred forty-nine

k) six hundred twenty-one trillion, one hundred fourteen billion, nine hundred eighty-seven million, two thousand, five hundred eighty-one

l) 172,215,702,301

m) 7,000,406,347,015

n) 7,814,000,000

o) 7,813,560,000

p) 8,000,000,000

q) 7,813,600,000

1.1b

a) 13, 17, 61

b) 16, 102, 2434, 4, 16, 156

c) 231, 6033, 42, 63, 99

d) 15, 7855, 239670, 65, 8000

e) 240, 1212, 6954

f) 1000, 5020, 84750, 90, 21054780

g) $165: 3 \cdot 5 \cdot 11$

h) $2358: 2 \cdot 3 \cdot 3 \cdot 131$

i) $81: 3 \cdot 3 \cdot 3 \cdot 3$

j) $64: 2 \cdot 2 \cdot 2 \cdot 2 \cdot 2 \cdot 2$

k) $125: 5 \cdot 5 \cdot 5$

l) 60

m) 105

n) $3 \cdot 2 \cdot 2 \cdot 5 = 60$

o) $3 \cdot 5 \cdot 7 = 105$

p) $2 \cdot 2 \cdot 2 \cdot 2 \cdot 2 \cdot 2 \cdot 3 \cdot 7 = 1344$

1.2a

a) $5y + 15 = 20; \frac{r-2}{r+2} = 4; (x - 2)(x + 5) = 12$

b) 81

c) 144

d) 125

e) -8

f) Seventeen is greater than two

g) x is greater than or equal to 8

h) negative five is less than seven

i) x is less than or equal to two

j) six more than x; six added to x

k) two subtracted from x; two less than x

l) fifteen more than x; or x plus 15

m) seven less than y

n) The sum of four x-squared and fifteen

o) The difference of seventeen x and four y

p) The sum of nine x-cubed and six

q) The difference of two x-squared and three y-squared

r) The sum of five x-squared and twenty-one

s) The difference of five x-squared and twenty-one

t) The product of eight and x

u) The quotient of eight and x

v) The product of six and x-squared

w) The quotient of x-squared and six

x) The product of twelve and x

y) The quotient of x and twelve

z) The quotient of twelve and x

aa) Three times the sum of x and y

bb) The sum of three times x and three time y

cc) Two times the difference of r and s

dd) The difference of two times r and two times s

ee) Twelve times the sum of a and b; or the product of 12 and the sum of a and b.

ff) The difference of twelve times a and twelve times b

1.2b

a) 49

b) 14

c) 34

d) 46

e) 36

f) -6

g) 49

h) 99

i) 43

1.2c

a) 30

b) a

c) c

d) c

e) c

f) 14x

g) 13x

h) 6x+10y

i) 26x-5y

j) b

k) $17x^2 - 3x$

l) $8x^2 + 2y^2 + 7y + 3x$

m) 11xy

n) $14x^2y$

o) $21x^2y + 1xy^2$

p) $5x^2 + 7y^2 + 4y$

q) $6x^3, -15x^2, x, 9$

r) 6, −15, 1, 9

1.3a

a) <

b) >

c) <

d) <

e) >

f) -6

g)-101

h)>

i) 25

j) 8

k) 21

l) 17

m) 17

n) 3

o) 52

p) 258

q) 9

r) 9

s) -8

t) -8

u) 21

v) 0

w) 37

1.3b

a) -2

b) -14

c) 3

d) 9

e) 14

f) -20

g) -11

h) -12

i) -21

j) 17

k) 7

l) -8

m) -11

n) -5

o) 14

p) -2

q) -9

r) 6

s) -8

t) 10

u) 12

v) -1

w) 19

x) 3

y) 11

z) 1

aa) -44

bb) -29

cc) 6

dd) 8

ee) -26

ff) 1

gg) 4

hh) -10

ii) 21

1.4a

a) 10

b) 10

c) 7

d) 7

e) 6

f) 6

g) 6

h) No

i) 24

j) 63

k) 66

l) positive; -55

m) -42

n) -24

o) -8

p) negative

q) -9

r) -5

s) 6

t) -5

u) 9

v) 64

w) positive

x) -32

y) negative

z) -27

aa) 25

bb) 256

cc) -216

dd) 15

ee) -23

ff) –3

gg) 11

hh) 25

ii) -18

jj) 25

kk) -125

1.4b

a) 17+(-2)

b) 6 · 15

c) $\frac{36}{-6}$

d) 27-5

e) 12-9

f) 80-45

g) 64-32

h) $[13 + (-17)] + 2$

i) $[-4 + (-7)] + 2$

j) (12-3)+11

k) $[19 + (-2)] - 10$

l) $(3 \cdot 2) + 5$

m) $(-4) \cdot 8 - 12$

n) $\left(\frac{42}{7}\right) + 5$

o) $\left(\frac{72}{-9}\right) - 3$

p) 70+15

q) 22-14

r) $-2 - (-8)$

s) $5 \cdot 10$

t) $\frac{65}{5}$

1.5a

a) $\frac{4}{12}$

b) $\frac{3}{9}$

c) b

d) $\frac{4}{6}, \frac{6}{9}, \frac{8}{12}$

e) d

f) $\frac{3}{7}$

g) $\frac{3}{4}$

h) $\frac{5}{6}$

i) $-\frac{5}{6}$

j) $-\frac{2}{3}$

k) $\frac{a}{b}$

l) $\frac{x}{2y}$

m) 4

n) 10

o) $\frac{1}{6}$

p) $\frac{1}{20}$

q) $\frac{1}{9}$

r) $\frac{6}{25}$

s) $-\frac{6}{28}$

t) $\frac{1}{6}$

u) $\frac{y}{42}$

v) $\frac{2x}{50}$

w) $\frac{5}{6}$

x) $\frac{5}{4}$

y) $-\frac{10}{7}$

z) $\frac{1}{2}$

aa) $\frac{33}{32}$

bb) $\frac{7x}{15}$

cc) $\frac{15}{8y}$

dd) $\frac{y}{2}$

1.5b

a) $\frac{8}{21}$

b) $\frac{9}{10}$

c) $\frac{5y}{6}$

d) $\frac{2x}{3}$

e) $\frac{3x}{2}$

f) $\frac{3}{2x}$

g) $-\frac{1}{6}$

h) 4

i) $\frac{27}{2}$

j) $\frac{x}{y}$

k) $\frac{(x-2)}{y}$

l) $\frac{(x+y)}{z}$

1.6a

a) $\frac{5}{7}$

b) $\frac{5}{12}$

c) $\frac{1}{2}$

d) $\frac{5}{11}$

e) $\frac{13}{16}$

f) $\frac{2}{3}$

g) $-\frac{1}{4}$

h) $\frac{y+3}{5}$

i) $\frac{7+r}{15}$

j) $\frac{x-1}{3}$

k) $\frac{z-2}{7}$

l) The 6 would need the 2 · 2. The 8 would need the 3.

m) $\frac{10}{42}$

n) $\frac{11}{10}$

o) $\frac{47}{40}$

p) $\frac{2}{21}$

q) $\frac{7}{24}$

r) $-\frac{8}{15}$

s) $\frac{7y+3}{42}$

t) $\frac{3+2x}{10}$

u) $\frac{5z+12}{40}$

v) $\frac{7x-3}{42}$

w) $\frac{10y-3}{24}$

x) $\frac{1-9z}{15}$

1.6b

a) $\frac{1}{18}$

b) $\frac{5}{6}$

c) $\frac{7}{12}$

d) $\frac{10}{7}$

e) $\frac{13}{15}$

f) $\frac{4}{5}$

g) $\frac{13}{12}$

h) $\frac{1}{14}$

i) $\frac{32}{20}$

j) $\frac{10}{217}$

k) $\frac{13}{12}$

l) $\frac{23}{21}$

m) $-\frac{23}{18}$

n) $\frac{1}{12}$

o) $\frac{1}{9}$

p) $-\frac{3}{28}$

q) $\frac{9}{64}$

1.7a

a) 1

b)

Hundred Thousand	Ten Thousand	Thousand	Hundred	Tens	Ones		Tenths	Hundredths	Thousandths	Ten-Thousandth	Hundred-Thousandth
		5	7	8	2	.	1	4	9	2	

c) Hundred

d) Hundredths

e) Thousands

f) Thousandths

g) one hundred twenty-five and nine tenths

h) three and three tenths

i) sixteen hundredths

j) four hundred twelve thousandths

k) seven hundredths

l) sixty-seven and twelve hundredths

m) four hundred seventy-five and eighteen hundredths

n) two thousand forty-five and three thousandths

o) six and one hundredth

p) 412.015

q) 5.25

r) 2000.6

s) 699.002

t) 631.45

u) 631.4

v) 600

w) 631

x) 51.653

y) 14.939

1.7b

a) 653.25

b) 16389.69252

c) 48769.85

d) 487698.5

e) 4876985

f) 7.761

g) 9.782

h) -5.455

1.7c

a) $\frac{4}{5}$

b) $\frac{1}{4}$

c) $\frac{37}{100}$

d) $\frac{1474}{10000}$

e) .8

f) .875

g) $.\overline{3}$

h) $.\overline{18}$

i) 2.375

j) .68

k) .16

l) 1.15

m) 12%

n) 65.7%

o) 11.17%

p) .3%

1.8a

a) 5

b) 6

c) 11

d) 13

e) -7

f) -8

g) $\frac{121}{1}$

h) $\frac{12}{1000}$

i) $-\frac{6}{1}$

j) $\frac{8}{10}$

k) Rational: $4.\overline{125}, 6.25$

Irrational: 2.71828..., 7.1542...

l) Rational: $\sqrt{4}, \sqrt{225}$

Irrational: $\sqrt{48}, \sqrt{35}$

m) d, f

n) e, f

o) d, f

p) c, d, f

q) e, f

r) a, b, c, d, f

s) a, b, c, d, f

t) b, c, d, f

u) None

1.8b

a) 4.4

b) 3:7

c) 1

d) <

e) >

f) >

g) -4

h) -3.1

i) -.8

j) >

k) <

l) <

m) >

n) <

o) >

p) <

q) >

r) <

s)

t)

u)

v)

w)
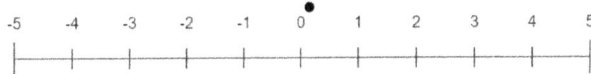

x)

y)

z)

aa)

bb)
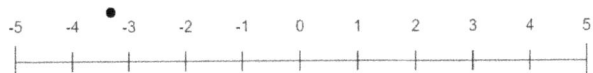

<u>1.9a</u>

a) 19

b) 19

c) 7

d) 7

e) No

f) 10v+29u

g) -7r+24s

h) 30

i) 30

j) -21

k) -21

l) No

m) 9

n) 9

o) 4

p) 4

q) No

r) 3y

s) 8v

t) 24

u) 24

v) -90

w) -90

x) No

y) 8

z) -15

aa) 1254

bb) # remains the same

cc) 5x

dd) -17y

ee) 3

ff) -7

gg) 57

hh) # remains the same

ii) 16x

jj) 8y

kk) 0

ll) 0

mm) 0

nn) You get zero

oo) 0

pp) 0

qq) $\frac{1}{12}$

rr) $\frac{1}{9}$

ss) 2

tt) 1

uu) 1

vv) 1

ww) 1

xx) You get 1

yy) 1

zz) 1

aaa) o

bbb) 0

ccc) 0

ddd) You get zero

eee) 0

fff) 0

ggg) 0

hhh) 0

iii) 0

jjj) 0

kkk) 0

lll) You get zero

mmm) 0

nnn) 0

ooo) undefined

ppp) undefined

qqq) undefined

rrr) undefined

sss) undefined

ttt) undefined

1.9b

a) 15

b) 15

c) 15

d) 2x+10

e) 10x-20

f) -3x-20

g) 2x+xy

h) -3x-4

i) -2y+15

j) 7+4x

k) -20-6x

l) 23x-18

m) -18y-6

n) -24x-24

o) 4y+20

<u>1.10a</u>

a) $\frac{2}{5}$

b) $\frac{2}{7}$

c) $\frac{1 \, yard}{3 \, feet}$ or $\frac{3 \, feet}{1 \, yard}$

d) 6 feet

e) 108 inches

f) 1.575 tons

g) 26400 feet

h) 316800 inches

i) $.\overline{5}$ hrs

j) 128 tablespoons

k) 1200 mm

l) 1.15 meters

m) 3200000 mm

n) 13 lbs 15 ounces

o) 18 feet 3 inches

p) 15 min 10 sec

q) 11 feet 3 inches

r) 9 lbs 4 ounces

s) 4 hours 45 minutes

<u>1.10b</u>

a) 3.3 meers

b) .8 meters

c) 5.2 meters

d) .6 liters

e) 1.8 kg

f) 9.786 meters

g) 12.15 kgs

h) 1.54 lbs

i) 10.65 feet

j) 12.78 degrees Celsius

k) 59 degrees Fahrenheit

<u>2.1a</u>

a) $3y = 21; 9x = 81$

b) y=7

c) r=15

d) c=12

e) y=10

f) s=32

g) a=28

h) y=-18

i) t=-5

j) b=-24

k) $y = \frac{1}{6}$

l) $r = \frac{5}{8}$

m) c=7.2

n) d=10.5

o) x=-7.1

p) y=14

q) z=4

r) x=-7

s) s=-3

t) x=13

u) x=7

v) y=-34

w) v=8

x) x=-28

y) y=6

2.1b

a) x=6

b) y+21=43

y=22

c) x-3=12

x=15

d) y-14=30

y=44

e) x+9=15

x=6

f) 12y-11y=9

y=9

g) 110x-109x=-32

x=-32

h) x+10=24

x=14

i) 65+x=80

x=15

j) x+90=150

x=60

k) x-2.75=5.50

x=8.25

l) x-1575=21450

x=23025

2.2a

a) 7

b) 4

c) 9

d) 6

e) -9

f) -6

g) 6

h) 13

i) 18

j) 14

k) -4

l) 18

m) 30

n) 75

o) 208

p) -32

q) 242

r) -21

s) 7

t) $\frac{10}{9}$

u) $\frac{1}{6}$

v) -1

w) $\frac{4}{5}$

x) 6

y) 10

z) 7

aa) -1

bb) -11

cc) 5

2.2b

a) 13y=156

y=12

b) $\frac{y}{11} = 20$

y=220

c) $\frac{n}{4} = -8$

n=-32

d) $\frac{x}{7} = 15$

x=105

e) $\frac{y}{-4} = 14$

y=-56

f) $\frac{1}{4}x = 23$

$x = 92$

g) $\frac{3}{4}x = 11$

$x = \frac{44}{3}$

h) $\frac{1}{3} + x = \frac{13}{12}$

$x = \frac{3}{4}$

i) $y - \frac{3}{4} = \frac{31}{20}$

$y = \frac{23}{10}$

j) x=7

k) x=$15.75

l) x=$177.50

2.3a

a) y=7

b) z=5

c) r=6

d) x=-8

e) y=10

f) y=6

g) r=-3

h) x=7

i) r=4

j) z=-6

k) x=7

l) x=1

m) c=-6

n) w=-2

o) x=-17

2.3b

a) y=7

b) z=5

c) v=-2

d) x=7

e) $x = -\frac{20}{7}$

f) $r = \frac{19}{4}$

g) x=32

h) $y = \frac{36}{5}$

i) x=10

j) y=-160

2.4a

a) r=3

b) x=4

c) x=2

d) y=30

e) $r = -\frac{2}{3}$

f) x=2

2.4b

a) False

b) True

c) True

d) False

e) 0=0 True

f) 8=0 False

g) -2=8 False; Contradiction

h) x=-3 Conditional

i) 0=0 True; Identity

2.5a

a) $\frac{5}{6}$

b) $\frac{11}{12}$

c) Common Denominator

d) x=1

e) y=1

f) x=2

g) x=2

h) $y = -\frac{1}{4}$

i) y=15

j) $a = \frac{15}{2}$

k) x=-6

l) y=-14

m) x=3

2.5b

a) x=2

b) y=5

c) y=2

d) x=3

e) x=-31

f) y=5

2.6a

a) 55 miles per hour; 15 miles per hour

b) 120 miles

c) d=135 miles

d) r=50 mph

e) 62 mph

f) 70 mph

g) t=6 hours

h) t=8 hours

2.6b

a) $t = \frac{d}{r}$

b) $h = \frac{2A}{b}$

c) $P = \frac{I}{rt}$

d) $a = P - b - c$

e) $b = P - a - c$

f) y=-2x+5

g) y=-4x+9

h) $y = \frac{-2x+7}{5}$

i) $y = \frac{-7x+5}{9}$

2.7a

a) yes

b) yes

c) 5<15

d) yes

e) yes

f) 6>4

g) yes

h) yes

i) 6>4

j) yes

k) -16>-10

l) No

m) x>4 $(4, \infty)$

n) y<3 $(-\infty, 3)$

o) $c \geq -9$ $[-9, \infty)$

p) $x \leq 11$ $(-\infty, 11]$

q) $(9, \infty)$

r) $(-3, \infty)$

s) $[9, \infty)$

t) $[-3, \infty)$

u) $(-\infty, 9)$

v) $(-\infty, -3]$

2.7b

a) x<7 $(-\infty, 7)$

b) $y \leq 4$ $(-\infty, 4]$

c) $x > 3$ $(3, \infty)$

d) $r \leq -5$ $(-\infty, 5]$

e) -15>21 No Solution

f) $14 \geq -6$ $(-\infty, \infty)$

g) $x \leq -11$ $(-\infty, -11]$

3.1a

a) Hello. How are you? I like dogs.

b) I am from the United States. Where are you from?

c) x=$330

d) The original price was $330.

e)

Math Language	English
x	Number of Adult Tickets

f) x=125

g) The number of adult tickets sold was 125.

h)

Math Language	English
x	Number of Textbooks

x=5; The student has 5 textbooks.

i)

Math Language	English
x	Number of Tulips

x=7; The number of tulips was 7.

3.1b

a) x=63

b) x=5

c) x=11

d) x=25

e) x=21 (and 36)

f) x=-19 (and -13)

g) x=49 (and 44)

h) x=17 (and 38)

i) x=16 (and 50)

j) x=13 (and 14)

k) x=45 (and 46)

l) x=-23 (and -22)

m) x=32 (and 33 and 34)

n) x=-9 (and -8 and -7)

o) x=27 (and 29)

p) x=-43 (and -41)

q) Consecutive even needs to take two more steps

r) x=30 (and 32 and 34)

3.2a

a) x=20

b) x=5

c) 25%

d) 56.5%

e) 7.5

f) x=24.6

g) x=210

h) x=12.30

i) $21

j) x=18.5

k) 52.9%

3.2b

a) 40%

b) 12.73%

c) 6.25%

d) 10.53%

e) $3900

f) $3308.82

g) 1.15%

3.2c

a) 90%

b) 70%

c) 82%

d) 49.40

e) $15.60

f) 276.25

g) $48.75

h) 110%

i) 130%

j) 118%

k) $74.10

l) $9.10

m) $344.50

n) $19.50

o) 26%

p) $6.50

3.3a

a) x=8 (and 21)

b) $1.25

c) $1.20

d) $.35

e) x+3 Nickels; x Quarters

f) x=6

g) 6 quarters, 9 nickels

h) 7 nickels; 20 dimes

i) x=15

j) 15 nickels; 33 dimes

k) 1.8

l) 1.05

m) x=7; 7 .27 cent stamps; 19 .15 cent stamps

n) 225 adult; 466 child

o) 180 adult; 370 child

p) 157 adult; 515 child

3.3b

a) 2.4 lbs of raisins; 3.6 lbs of oats

b) $5\frac{1}{3}$ lbs of raisins; $2\frac{2}{3}$ lbs of oats

c) 4 lbs of cereal; 4 lbs of chocolate chips

d) $7500 in the 5% fund; $2500 in the 7% fund

e) $10,500 in the 3% fund: $4500 in the 8% fund

3.4a

a) 180 degrees

b) 41 degrees

c) 62 degrees

d) 55 degrees

e) 29 (and 61) degrees

f) 20 meters

g) 22 feet

h) 15 feet

l) 31.25 meters

3.4b

a) 13

b) 8

c) 12

d) 8.49

3.4c

a) length is 40 meters

b) 25 feet

c) 35 yards

d) 70 meters; 55 meters

e) width 10 feet; length 25 feet

f) 5 meters

g) 20 meters

3.5a

a) 165 miles

b) 165 miles

c) 385 miles

d) 385 miles

e) They were the same

f) They were the same

g) r=18 mph

h) Brother 18 mph; You 30 mph

i)

	Rate	Time	Distance
Standard	x	7	7x
Express	x+15	5.5	5.5(x+15)

j) x=55

k) Standard 55 mph; Express 70 mph

l) r=40

m) First 40 mph; Second 55 mph

n) First 34 mph; Second 44 mph

o) time is 3.5 hours

p) time is 3.5 hours

3.5b

a) 180 miles

b) 30 miles

c) 120 miles

d) 90 miles

e) .33 hours

f) .75 hours

g) .5 hours

h) r=5

i) your brother: 5 mph; you: 10 mph

j) brother: 12 mph; you: 18 mph

k) r=10 mph

l) swim: 10 mph; bike 30 mph

m) slow: 15 mph; faster: 60 mph

3.6

a) $\leq$

b) >

c) <

d) $\leq$

e) >

f) $\geq$

g) $x \leq 18.5; 18\ souvenirs$

h) $x \geq 5.5; 6\ cars$

i) $x \geq 13$; 13 *televisions*

j) $x \geq 28.46$; 29 *lawns*

k) $x \geq 98.09$; 99 *hours*

4.1a

a)
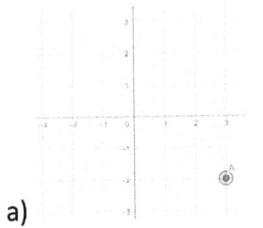

b)
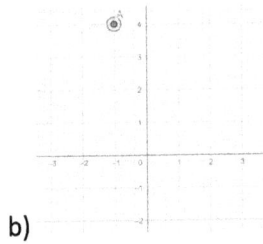

c)
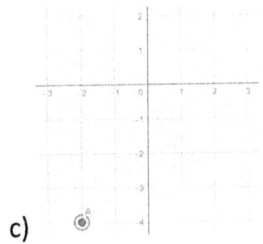

d) Quadrant II
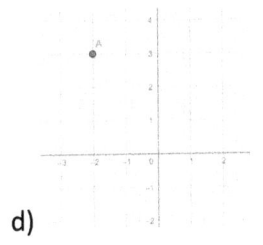

e) Quadrant I
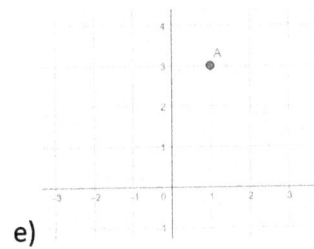

f)
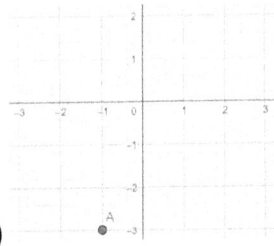
Quadrant III

g) x-axis

h) y-axis

i) x-axis

j) x-coordinate: -2

y-coordinate: 3

k) x-coordinate: -1

y-coordinate: -3

l) x-coordinate: 8

y-coordinate: 12

m) x-coordinate: 0

y-coordinate: 0

n) origin

4.1b

a)

x	y
3	5
5	3
2	6
-1	9
-4	12

b)
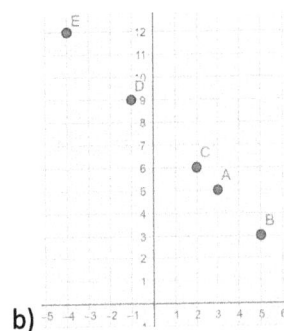

x	y
3	2
2	4
5	-2
1	6
-1	10

c)

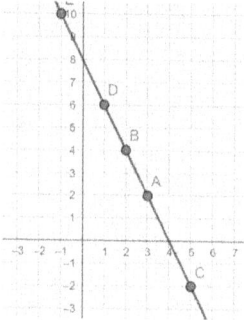

d)

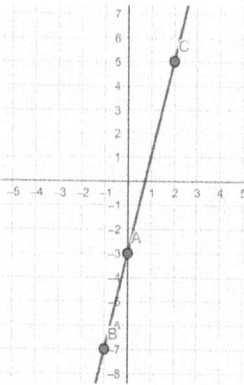

e)

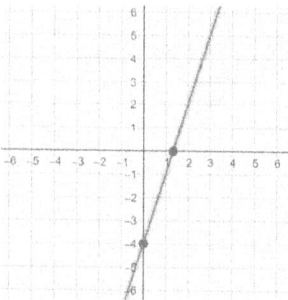

f)

g) a, c, d

4.2a

a) a, c

b)

x	y
4	-2
0	-3
-4	-4

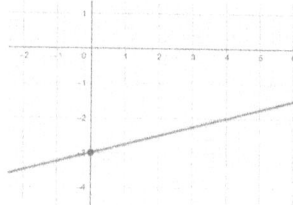

c)

x	y
5	1
0	-2
-5	-5

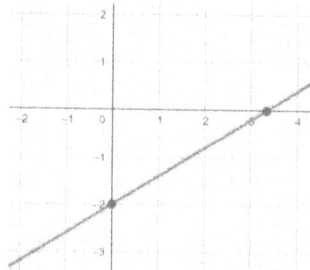

4.2b

a)

x	y
1	3
0	3
-1	3

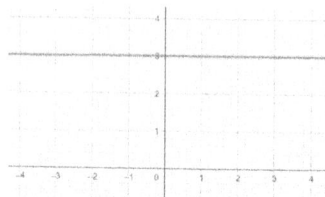

b)

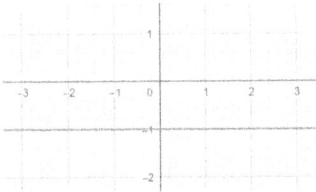

c)

x	y
-2	-1
-2	0
-2	1

d)

e) horizontal

f) vertical

4.3

a) x-intercept: (-1, 0)

y-intercept: (0, 2)

b) x-intercept: (6, 0)

y-intercept: (0, -3)

c) x-intercept: (2, 0)

y-intercept: (0, 3)

d)

e) x-intercept: (2, 0)

y-intercept: (0, -5)

f)

4.4a

a) 2 up; 3 right

b) 2 up; 3 right

c) the slope $\frac{2}{3}$

d) 2 down; 3 left

e) the negatives cancel

f) down 3

g) right 4

h) yes

i) up 3

j) left 4

k) $\frac{-3}{4}, \frac{3}{-4}$

l) yes

m) down 6

n) right 8

o) $-\dfrac{6}{8}$

p) $-\dfrac{6}{8} = -\dfrac{3}{4}$

q) $-\dfrac{2}{1}$

r) $\dfrac{2}{5}$

s) $\dfrac{1}{4}$

t) $-\dfrac{1}{4}$

u) positive: $y = \dfrac{3}{2}x - 1$; $y = \dfrac{5}{7}x + 6$

negative: $y = -\dfrac{1}{5}x + 2$; $y = -2x + 3$

v)

w) $\dfrac{1}{4}$

x) 1

y) 4

z) $\dfrac{1}{4}$

aa) $\dfrac{3}{2}$

bb) $-\dfrac{2}{3}$

cc) $\dfrac{-2}{2} = -1$

dd) 0

ee) the y values are always the same

ff) undefined

gg) because the x coordinates are in the denominator and they are always the same.

4.4b

a)

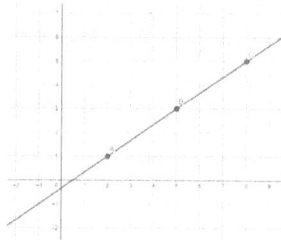

b)

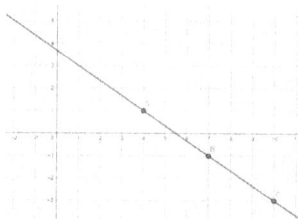

c)

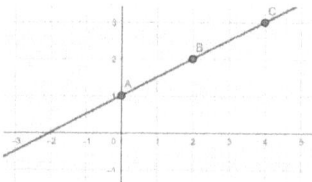

d)

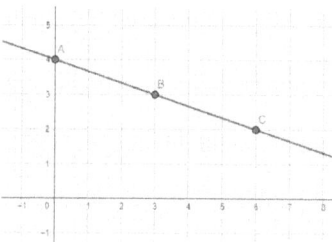

e)

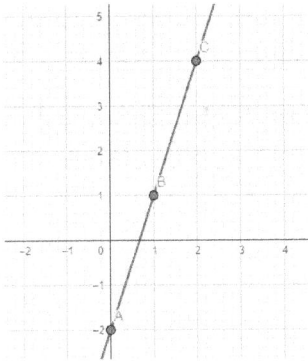

f)

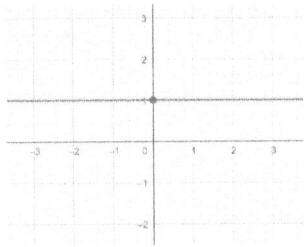

4.5a

a) (0, 5)

b) (0, 1)

c) (0. -3)

d) The number alone is the intercept.

e) $-\frac{3}{2}$; (0, 5)

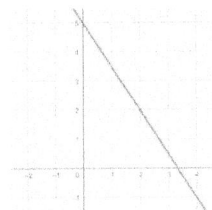

f) $\frac{4}{5}$; (0, -4)

g) $-\frac{1}{3}$; (0, 7)

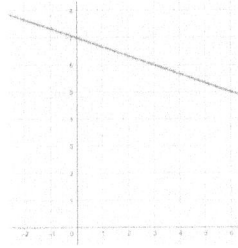

h) $y = 2x + 3$

i) $y = 5x + 8$

j) $y = -3x + 2$

k) $y = -2x + 3$

l) $y = -\frac{2}{3}x + 4$

m) $y = -\frac{4}{3}x + 2$

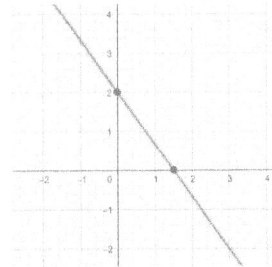

n) $y = \frac{3}{2}x - 2$

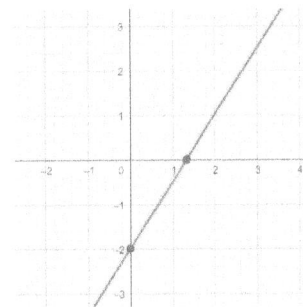

4.5b

a) $-\dfrac{2}{3}$

b) (0, 3)

c)

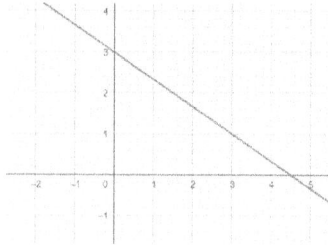

d) (6, 0)

e) (0, 3)

f)

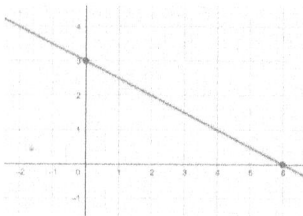

g) vertical

h) horizontal

i)

j)

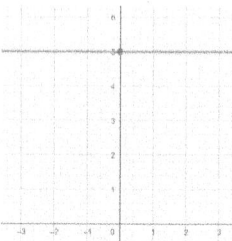

k) 35 months old

l) As the score of a child goes up 4 points, we expect the age of the child to go up 3 months.

m) As the score of a child goes up 1 point, we expect the age of the child to go up 2 months.

4.5c

a)

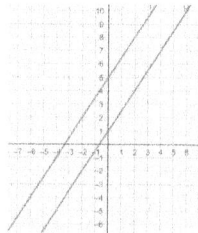

b) parallel

c) the same slope

d)

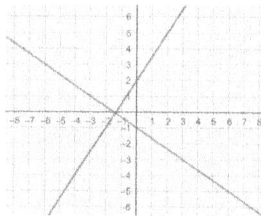

e) perpendicular

f) inverse slopes (upside down) with the opposite sign

g) neither

h) parallel

i) perpendicular

4.6a

a) $\frac{4}{3}$

b) $(0, -2)$

c) $y = \frac{4}{3}x - 2$

d) slope formula

e) $(6, 6)$

f) $\frac{4}{3}$

g) $y - 6 = \frac{4}{3}(x - 8)$

h) $y = \frac{4}{3}x - 2$

i) yes

j) $y = \frac{2}{3}x - 4$

k) $y = -\frac{1}{3}x - 2$

l) $(6, 6)$

m) $(3, 2)$

n) $\frac{4}{3}$

o) $y = \frac{4}{3}x - 2$

p) $y = \frac{4}{3}x - 2$

q) $y = 2x - 4$

r) x = 5

s) y = -1

4.6b

a) $\frac{4}{3}$

b) It is the same.

c) $\frac{4}{3}$

d)

e) $-\frac{2}{3}$

f)

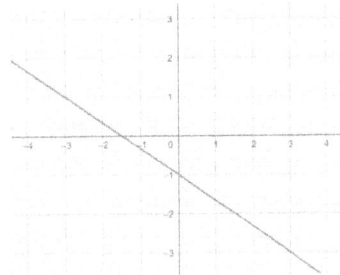

g) $\frac{1}{2}$

h) $y = \frac{1}{2}x$

i) $-\frac{2}{3}$

j) it is the inverse with the opposite sign

k) $\frac{3}{2}$

l)

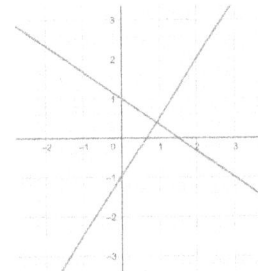

m) $\frac{1}{3}$

601

n)

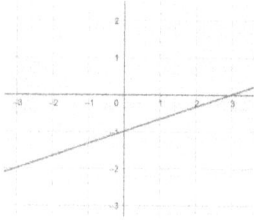

o) -2

p) $y = -2x + 5$

4.7

a)

	Coordinates	True/False
Point A	(-2, 2)	TRUE
Point B	(-4, 1)	TRUE
Point C	(4, 0)	FALSE
Point D	(2, -4)	FALSE
Point E	(0, 2)	FALSE

b) It is on the dashed line but not true.

c) $y \geq \frac{3}{2}x + 2$

d)

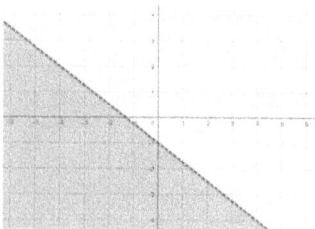

e)

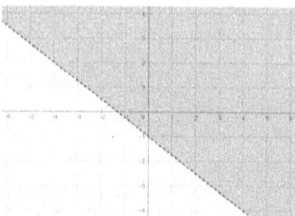

f)

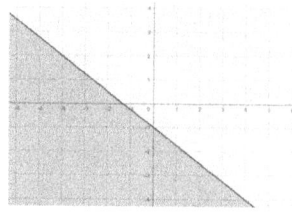

g)

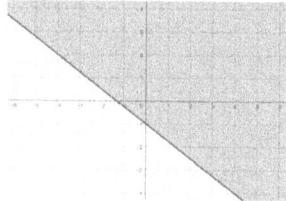

h) >, <

i) don't include the points on the line.

j)

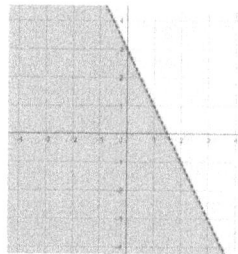

k)

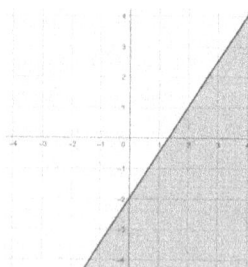

l)

5.1

a) (3, 1)

b) (1, -2)

c) (-5, -4)

d)

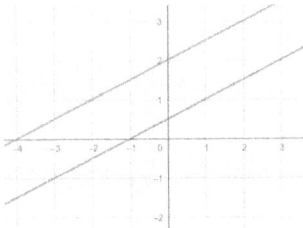

e) the lines don't intersect.

f) the lines are parallel.

g) (-3,1) is a solution

5.2

a) y=15

b) x=50

c) (-2, 1)

d) (-2, -4)

e) (4, -1)

f) No solution. They don't intersect.

g) The lines are parallel.

h) They are the same line. Infinite solutions.

i) Infinite solutions because they are touching everywhere.

5.3

a) 7x

b) 5x=20

c) x=4

d) y=-1

e) (4, -1)

f) (-5, -4)

g) (0, -3)

h) (1, 3)

i) infinite solutions

5.4a

a) x=54, y=65

b) x=27; y=15

c) 40x+15y=750

d) 15 calories per minute on the elliptical, 10 calories per minute on the bike

e) 51 degrees and 39 degrees

f) The two angles are 106 degrees and 74 degrees

g) The length will be 70 feet; the width will be 50 feet

h) The length will be 110 feet; the width will be 60 feet

5.4b

a) First friend 12 hours; second friend 11 hours

b) The speed of the plane is 318.75 mph; the speed of the wind is 81.25 mph

c) The speed of the boat is 22.5 mph; the speed of the water current is 2.5 mph

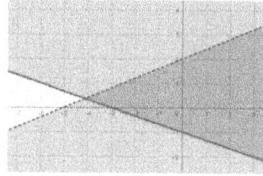

5.5a

a) 375 adult tickets; 875 children's tickets

b) 780 adult tickets; 770 children's tickets

c) 26 nickels; 57 dimes

d) 23 dimes; 41 quarters

e) 12 lbs of cereal; 3 lbs of raisins

f) 4 lbs of cashews; 4 lbs of almonds

g) 12 ml; 19.25 ml

h) 48 ml of 10%; 32 ml of 35%

i) 52.5 ml of 35%; 17.5 ml of 55%

h)

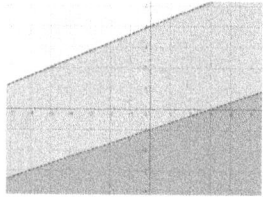

i)

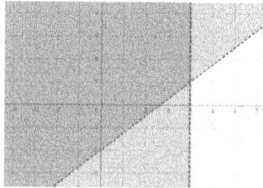

5.5b

a) $31250 in 3.5%; $18750 in 7.5%

b) $11250 in 4% CD; $18750 in 8% stock

c) $7500 in 8.5%; $1300 in 12.5%

d) $15000 at 8%; $17500 in 13%

5.6b

a) b, d

b) $x + y \geq 10$

c) $6x + 5.5y \leq 50$

d) c, d

5.6a

a) a, b

b) on dashed line

c) (-4, -2)

d) yes

e) (-3, 2)

f) No

6.1a

a) binomial

b) polynomial

c) monomial

d) trinomial

e) seven

f) four

g) six

h) two (second)

i) $12x^2 + 3x + 17$

j) $4y^3 + 13y^2 + 6y + 1$

k) $5x^2y + 3xy^2 + 15xy + 1$

l) $-2x^2 - 5x + 6$

m) $-2x^2 - 5x + 6$

n) $4x^2 + 23x - 29$

o) $4y^3 - 13y^2 - 8y - 15$

p) $-x^2y + 3xy^2 - 5xy + 11$

<u>6.1b</u>

a) -45

b) 43

c) 50

d) 43

e) 10

f) 2

g) -16

h) 0

<u>6.2a</u>

a) b

b) a

c) y^2

d) v^6

e) b

f) x^9

g) y^9

h) x^4

i) y^{10}

j) a

k) x^6

l) y^{12}

m) x^{10}

n) a^{40}

o) c

p) x^6y^4

q) $m^{15}n^{12}$

r) $x^6y^{12}z^2$

s) $27x^6$

t) $16m^4n^6$

<u>6.2b</u>

a) x^4

b) y^5

c) a^{13}

d) $24x^8$

e) $50y^8$

f) $14x^9$

g) $24y^6$

h) $-24x^3$

i) $15x^2$

j) 3x-15

k) $10x^2 + 12x - 4$

l) 24xy+36

m) $50x^2 - 15x$

n) $8y^3 + 14y^2$

o) $24a^3 - 6a^2 + 15a$

p) $36x^2y^2 - 8xy$

q) $14x^4y^2 + 7x^3y - 7x^2y$

r) $-2x^4y - 10x^3y^2 - 2x^2y^3$

6.3

a) $12x^2 - 60x$

b) $4x^4 - 8x^3 + 12x^2$

c) $x^2 + 3x$

d) $2x + 6$

e) $x^2 + 5x + 6$

f) $6x^2 + 19x + 10$

g) $50y^2 - 8$

h) $7a^3 + 3a^2 - 7a - 3$

i) $6a^2b^2 - ab - 1$

j) $x^2 - 9$

k) $2x^3 + 4x^2 - 2x$

l) $10x^2 + 20x - 10$

m) $2x^3 + 14x^2 + 18x - 10$

n) $x^3 + 8x^2 + 16x + 3$

o) $x^3 + 2x^2 - 4x + 1$

6.4a

a) $x^2 + 6x + 9$

b) $y^2 - 4y + 4$

c) $4x^2 + 10x + 25$

d) They are the same

e) $4x^2 - 12x + 9$

6.4b

f) $x^4 - 2x^2 + 1$

g) $x^2 - 12x + 36$

h) $4x^2 - 12x^2y^2 + 9y^2$

i) $4m^4 + 32m^2 + 64$

a) $x^2 - 9$

b) $y^2 - 4$

c) $4x^2 - 25$

d) same but opposite signs in the middle

e) always negative

f) $4x^2 - 9$

g) $x^4 - 1$

h) $4x^2 - 9y^2$

i) $4m^4 - 64$

j) $(x - 9)$

k) $(2y + 7)$

l) $a^2 - b^2$

6.5a

a) x^3

b) y

c) a^5

d) subtraction

e) $\frac{1}{x^4}$

f) $\frac{1}{y^3}$

g) 1

h) 1